Charles Augustus Young

Uranography

A Brief Description of the Constellations Visible in the United States

Charles Augustus Young

Uranography
A Brief Description of the Constellations Visible in the United States

ISBN/EAN: 9783337186029

Printed in Europe, USA, Canada, Australia, Japan

Cover: Foto ©berggeist007 / pixelio.de

More available books at **www.hansebooks.com**

URANOGRAPHY

A BRIEF DESCRIPTION OF

THE CONSTELLATIONS VISIBLE IN THE UNITED STATES

WITH

STAR-MAPS, AND LISTS OF OBJECTS OBSERVABLE
WITH A SMALL TELESCOPE

BY

C. A. YOUNG, Ph.D., LL.D.

LATE PROFESSOR OF ASTRONOMY IN PRINCETON UNIVERSITY

---

*A SUPPLEMENT TO THE AUTHOR'S "ELEMENTS OF ASTRONOMY
FOR HIGH SCHOOLS AND ACADEMIES"*

GINN & COMPANY

BOSTON · NEW YORK · CHICAGO · LONDON

PREFACE.

THIS brief description of the constellations was prepared, at the suggestion of a number of teachers, as an integral part of the author's "Elements of Astronomy." It has been thought best, however, for various reasons, to put it into such a form that it can be issued separately, and used if desired in connection with the larger "General Astronomy," or with any other text-book. Since the Uranography also has to be used more or less in the open air at night, many will probably prefer to have it by itself, so that its use need not involve such an exposure of the rest of the text-book. All references marked Astr. are to the articles of the "Elements of Astronomy."

ALPHABETICAL LIST OF THE CONSTELLATIONS DESCRIBED OR MENTIONED IN THE URANOGRAPHY.

THE GREEK ALPHABET.

Letters.	Name.	Letters.	Name.	Letters.	Name.
Λ, α,	Alpha.	I, ι,	Iota.	P, ρ, ϱ,	Rho.
B, β,	Beta.	K, κ,	Kappa.	Σ, σ, ς,	Sigma.
Γ, γ,	Gamma.	Λ, λ,	Lambda.	T, τ,	Tau.
Δ δ,	Delta.	M, μ,	Mu.	Y, υ,	Upsilon
E, ε,	Epsilon.	N, ν,	Nu.	Φ, φ,	Phi.
Z, ζ,	Zeta.	Ξ, ξ,	Xi.	X, χ,	Chi.
H, η,	Eta.	O, ο,	Omicron.	Ψ, ψ,	Psi.
Θ, θ, ϑ,	Theta.	Π, π,	Pi.	Ω, ω,	Omega.

URANOGRAPHY,

OR

A Description of the Constellations.

———o⦂⦂⦂oo———

1. A general knowledge of the constellations sufficient to enable one to recognize readily the more conspicuous stars and their principal configurations, is a very desirable accomplishment, and not difficult to attain. It requires of course the actual study of the sky for a number of evenings in different parts of the year; and the study of the sky itself must be supplemented by continual reference to a celestial globe or star-map, in order to identify the stars observed and fix their designations. A well-made globe of sufficient size is the best possible help, because it represents things wholly without distortion, and is easily "rectified" (Astr. 528[1]) for any given hour, so that the stars will all be found in the proper quarter of the (artificial) heavens, and in their true relations. But a globe is clumsy, inconvenient out of doors, and liable to damage; and a good star-map properly used will be found but little inferior in efficiency, and much more manageable.

2. Star-Maps. — Such maps are made on various systems, each presenting its own advantages. None are without more or less distortion, especially near the margin, though they

---

[1] The references are to the articles in the Author's "Elements of Astronomy," to which this Uranography is a supplement.

differ greatly in this respect. In all of them the heavens are represented *as seen from the inside*, and not as on the globe, which represents the sky as if seen from the *outside; i.e.*, the top of the map is *north*, and the east is at the *left* hand; so that if the observer faces the south and holds up the map before and above him, the constellations which are near the meridian will be pretty truly represented.

3. We give a series of four small maps which, though hardly on a large enough scale to answer as a satisfactory celestial atlas, are quite sufficient to enable the student to trace out the constellations and identify the principal stars.

In the map of the *north* circumpolar regions (Map I.), the pole is in the centre, and at the circumference the right-ascension hours are numbered in the same direction as the figures upon a watch face; but with 24 hours instead of 12. The parallels of declination are represented by equidistant and concentric circles. On the three other rectangular maps, which show the equatorial belt of the heavens lying between + 50° and — 50° of declination, the parallels of declination are equidistant horizontal lines, while the hour-circles are vertical lines also equidistant, but spaced at a distance which is correct for declination 35°, and not at the equator. This keeps the distortion within reasonable bounds even near the margin of the map, and makes it very easy to lay off the place of any object for which the right ascension and declination are given.

The hours of right ascension are indicated on the central horizontal line, which is the equator, and at the top of the map are given the names of the months. *The word September, for instance, means that the stars which are directly under it upon the map will be near the meridian about nine o'clock in the evening during that month.*

4. The maps show all the stars down to the 4½ magnitude — all that are easily visible on a moonlight night. A few smaller stars are also inserted, where they mark some peculiar configuration or point out some interesting telescopic object. So far as practicable, *i.e.*, north of — 30° Declination, the magnitudes of Pickering's "Harvard Photometry" are used. The places of the stars are for 1900.

In the designation of clusters and nebulæ the letter M. stands for "Messier," who made the first catalogue of 103 such objects in 1784 ; *e.g.,* 97 M. designates No. 97 on that list. A few objects from Herschel's catalogue are denoted by ɰ with a number following.

The student or teacher who possesses a telescope is strongly urged to get Webb's "Celestial Objects for Common Telescopes." It is an invaluable accessory. (Longmans, Green & Co., N. Y.)

THE CIRCUMPOLAR CONSTELLATIONS.

We begin our study of Uranography with the constellations which are *circumpolar* (*i.e.,* within 40° of the north pole), because these are always visible in the United States, and so can be depended on to furnish land (or rather *sky*) -marks to aid in identifying and tracing out the others.

5. Ursa Major, the Great Bear (Map I.). — Of these circumpolar constellations none is more easily recognizable than Ursa Major. Assuming the time of observation as about eight o'clock in the evening on Sept. 22d (*i.e.,* 20^h sidereal time), it will be found below the pole and to the west. Hold the map so that the VIII. is at the bottom, and it will be rightly placed for the time assumed.

The familiar Dipper is sloping downward in the northwest, composed of seven stars, all of about the second magnitude excepting δ (at the junction of the handle to the bowl), which is of the third. The stars a and β are known as the "Pointers," because the line drawn from β through a, and produced about 30°, passes very near the Pole-star.

The dimensions of the Dipper furnish a convenient scale of angular measure. From a to β is 5°; a to δ is 10°; β to γ, 8°; from a to η at the extremity of the Dipper-handle (which is also the Bear's *tail*) is 26°.

6. The Dipper (known also in England as the "Plough," and as the "Wain," or wagon) comprises but a small part of the whole constellation. The head of the Bear, indicated by a scattered group of small stars, is nearly on the line from δ through α, carried on about 15°; at the time assumed (20ʰ sid. time), it is almost exactly below the pole. Three of the four paws are marked each by a pair of third or fourth magnitude stars 1½° or 2° apart. The three pairs are nearly equidistant, about 20° apart, and almost on a straight line parallel to the diagonal of the Dipper-bowl from α to γ, but some 20° south of it. Just now (20ʰ sid. time) they are all three very near the horizon for an observer in latitude 40°, but during the spring and summer they can be easily made out.

7. Names[1] of Principal Stars. —

α. Dubhe.	ε. Alioth.
β. Merak.	ζ. Mizar. The little star near it is
γ. Phecda.	Alcor, the "rider on his horse."
δ. Megrez.	η. Benetnasch or Alkaid.

Double Stars: (1) ζ (Mizar), Mags. 3 and 5; Pos.[2] 149°; Dist. 14″.5. In looking at this object the tyro will be apt to think that the small star shown by the telescope is identical with Alcor: a very low power eye-piece will correct the error. (Astr. Fig. 113.) The large star is itself a "spectroscopic binary" (see Art. 465*). (2) ξ, the southern one of the pair which marks the left hind paw. *Binary:* Mags. 4 and 5; Pos. (1910) (about) 120°, Dist. (about) 3″. Position and distance both change rapidly, the period being only 61 years. This was the first binary whose orbit was computed.

Clusters and Nebulæ: (1) 81 and 82 M., A.R. 9ʰ 45ᵐ, Dec. 69° 44′. Two nebulæ, one pretty bright, about half a degree apart. (2) 97 M., A.R. 11ʰ 07ᵐ, Dec. 55° 43′—2° south-following β. A planetary nebula.

---

[1] Capitals denote names that are *generally* used; the others are met with only rarely.

[2] The "position angle" of a double star is the angle which the line drawn from the larger star to the smaller one makes with the hour-circle. It is always reckoned from the north completely around through the east, as shown in Fig. A.

8. Ursa Minor, the Lesser Bear (Map I.). — The line of the
"Pointers" unmistakably marks out the Pole-star ("Polaris"
or "Cynosura"), a star of the second magnitude standing alone.
It is at the end of the tail of Ursa Minor, or at the extremity
of the handle of the "*Little* Dipper," for in Ursa Minor, also,
the seven principal stars form a dipper, though with the handle
bent in a different
way from that of
the other Dipper.
Beginning at "Po-
laris" a curved line
(concave towards
Ursa Major) drawn
through δ and ε
brings us to ζ, where
the handle joins the
bowl. Two bright
stars (second and
third magnitude), β
and γ, correspond to
the pointers in the
larger Dipper, and
are known as the
"Guardians of the
Pole": β is called

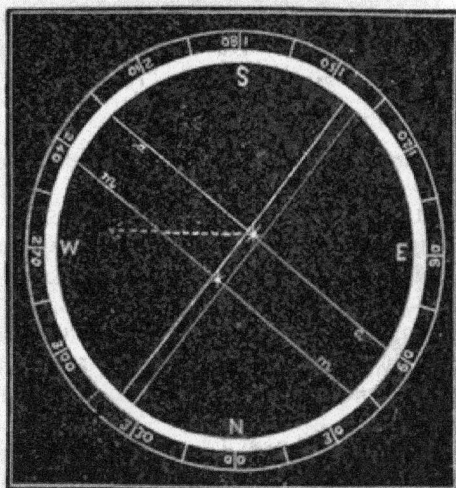

Fig. A. — Measurement of Distance and Position-Angle
of a Double Star.

"*Kochab.*" The remaining corner of the bowl is marked by
the faint star η with another still smaller one near it.

The Pole lies about 1¼° from the Pole-star, on the line joining
it to ζ Ursæ Majoris (at the bend in the handle of the large
Dipper).

Telescopic Object. Polaris has a companion of the 9½ magnitude,
distant 18″.6, — visible with a two-inch telescope.

9. Cassiopeia (Map I.). — This constellation lies on the
opposite side of the pole from the Dipper at about the same

distance as the "Pointers," and is easily recognized by the zigzag, "rail-fence" configuration of the five or six bright stars that mark it. With the help of the rather inconspicuous star κ, one can make out of them a pretty good *chair* with the feet turned away from the pole. But this is wrong. In the recognized figures of the constellation the lady sits with feet *towards* the pole, and the bright star a is in her bosom, while ζ and the other faint stars south of a, are in her head and uplifted arms: ι, on the line from δ to ϵ produced, is in the foot. The order of the principal stars is easily remembered by the word *Bagdei;* i.e., $\beta, a, \gamma, \delta, \epsilon, \iota$.

Names of Stars: a (which is slightly variable) is known as SCHEDIR; β is called CAPH.

Double Stars: (1) η, Mags. 4–7½. Large star orange; small one purple. Pos. 170° ±, Dist. 5″.5. Binary, with a period of some 200 years. Easily recognized by its position about half-way between a and γ, a little off the line. (2) ψ, A.R. 1ʰ 17ᵐ, Dec. 67° 21′; Triple; Mags. 4½, 9 and 9; Pos. A to (B + C) 106°, Dist. 29″; Pos. B–C 257°, Dist. 2″.9. Found on a line from η through γ produced three times the distance η–γ: rather difficult for a four-inch telescope.

10. The Sidereal Time determined by the Apparent Position of Cassiopeia. — The line from the Pole-star through Caph or β Cassiopeiæ (which is the *leader* of all the bright stars of the constellation in their daily motion) is almost exactly parallel to the Equinoctial Colure. When, therefore, this star is vertically *above* the Pole-star it is sidereal *noon;* it is 6ʰ when it is on the *great circle* (not the parallel of altitude) drawn from the Pole-star to the west point of the Horizon; 12ʰ when vertically below it; and 18ʰ when due east. A little practice will enable one to read the sidereal time from this celestial clock with an error not exceeding 15 or 20 minutes.

11. Cepheus (Map I.). — This constellation contains very few bright stars. At the assumed time (20ʰ sidereal) it is above and west of Cassiopeia, not having quite reached the meridian above the pole. A line carried from a Cassiopeiæ through β, and produced 20° (distance a . . . $\beta = 5°$ nearly) will pass very near to α Cephei, a star of the third magnitude, in the king's right shoulder.   β Cephei is about 8° due north of α, and γ about 12° from β, both also of third magnitude : γ is so placed that it is at the obtuse angle of a rather flat isosceles triangle of which β Cephei and the Pole-star form the two other corners.   Cepheus is represented as sitting behind Cassiopeia (his wife) with his feet upon the tail of the Little Bear, γ being in his left knee.   His head is marked by a little triangle of fourth magnitude stars, δ, ε, and ζ, of which δ is a remarkable variable with a period of 5½ days (see Astr. Table IV.). There are several other small variables in the same neighbor-hood, but none of them are shown on the map.

Names of Stars : α is *Alderamin*, and β is *Alphirk*.

Double Stars : (1) β, Mags. 3 and 8; Pos. 251°; Dist. 14″. (2) δ, Mags. larger star 3.7 to 5 (variable), smaller one 7; Pos. 192°, Dist. 41″; Colors, yellow and blue.   (3) κ, A.R. 20ʰ 13ᵐ, Dec. 77° 19′; Mags. 4.5 and 8.5; Pos. 124°; Dist. 7.″5; Colors, yellow and blue.

12. Draco (Map I.). — The constellation of Draco is char-acterized by a long, sinuous line of stars, mostly small, extend-ing half-way around the pole and separating the two Bears. A line from δ Cassiopeiæ drawn through β Cephei and ex-tended about as far again will fall upon the head of Draco, marked by an irregular quadrilateral of stars, two of which are of the 2½ and 3d magnitude.   These two bright stars about 4° apart are β and γ; the latter in its daily revolution passes almost exactly through the zenith of Greenwich, and it was by observations upon it that the aberration of light was discovered (Astr. 125).   The nose of Draco is marked by a smaller star, μ, some 5° beyond β, nearly on the line drawn through it from γ. From γ we trace the neck of Draco, eastward and downward[1] towards the Pole-star until we come to δ and ε and some smaller stars near them.   There the direction of the line is reversed,

---

[1] The description here applies strictly only at 20ʰ sid. time.

so that the body of the monster lies between its own head and
the bowl of the Little Dipper, and winds around this bowl until
the tip of the creature's tail is reached at the middle of the
line between the Pointers and the Pole-star. The constellation covers more than 12ʰ of right ascension.

13. One star deserves special notice: *a*, a star of the 3½
magnitude which lies half-way between Mizar (ζ Urs. Maj.)
and the Guards (β and γ Urs. Min.); 4700 years ago it was
the Pole-star, within 10′ or 15′ of the pole, and much nearer
than Polaris is at present, or ever will be. It is probable
that its brightness has considerably diminished within the
last 200 years; since among the ancient and mediæval astronomers it was always reckoned of the second magnitude.

Names of Stars: a is Thuban; β, *Alwaid;* and γ, *Etanin.*

Double Stars: (1) μ, Mags. 4 and 4½; Pos. 165°; Dist. 2″.5. Binary, with a probable period of about 600 years. (2) ε, Mags. 4,
8; Pos. 0°.0; Dist. 2″.9; yellow and blue. *Nebula,* A.R. 17ʰ 59ᵐ;
Dec. 66° 38. Planetary, like a star out of focus. This object is
almost exactly at the pole of the ecliptic, about midway between δ and
ζ Draconis, but a little nearer ζ.

14. Camelopardus (Map I.). — This is the only remaining one
of the strictly circumpolar constellations — a modern asterism containing no stars above fourth magnitude, and constituted by Hevelius
simply to cover the great empty space between Cassiopeia and Perseus
on one side, and Ursa Major and Draco on the other. The animal
stands on the head and shoulders of Auriga, and his head is between
the Pole-star and the tip of the tail of Draco.

The two constellations of Perseus (which at 20ʰ sidereal time is
some 20° below Cassiopeia) and of Auriga are partly circumpolar, but
on the whole can be more conveniently treated in connection with the
equatorial maps. Capella, the brightest star of Auriga, and next to
Vega and Arcturus the brightest star in the northern hemisphere, at
the time assumed (Sept. 22, 8 P.M.), is a few degrees above the horizon
in the N.E. Between it and the nose of Ursa Major is part of the
constellation of the Lynx, — a modern asterism made, like Camelopardus, merely to fill a gap.

15. The Milky Way in the Circumpolar Region. — The only circumpolar constellations traversed by it are Cassiopeia and Cepheus. It enters the circumpolar region from the constellation of Cygnus, which at 20^h sidereal time is just in the zenith, sweeps down across the head and shoulders of Cepheus, and on through Cassiopeia and Perseus to the northeastern horizon in Auriga. There is one very bright patch a degree or two north of β Cassiopeiæ; and half-way between Cassiopeia and Perseus there is another bright cloud in which is the famous cluster of the "Sword Handle of Perseus" — a beautiful object for even the smallest telescope.

---

16. Andromeda (Map II.). — Passing now to the equatorial maps and beginning with the northwestern corner of Map No. II., we come first to the constellation of Andromeda, which will be found exactly overhead in our latitudes about 10 o'clock in the middle of November, or at 8 o'clock a month later. Its characteristic configuration is the line of three second-magnitude stars, α, β, and γ, extending east and north from α, which itself forms the N.E. corner of the so-called "Great Square of Pegasus," and is sometimes lettered as δ Pegasi. This star may readily be found by extending an imaginary line from Polaris through β Cassiopeiæ, and producing it about as far again: α is in the head of Andromeda, β in her waist, and γ in the left foot. About half-way from α to β, a little south of the line, is δ (of the third magnitude) with π and ϵ of the fourth magnitude near it. A line drawn northwesterly from β nearly at right angles to the line $\beta\gamma$, will pass through μ at a distance of about 5°, and produced another 5° will strike the "great nebula" (Astr. 470), which forms a little obtuse-angled triangle with ν and a sixth-magnitude star known as 32 Andromedæ.

Andromeda has her mother, Cassiopeia, close by on the north, and at her feet is Perseus, her deliverer, while her head rests upon the shoulder of Pegasus, the winged horse which brought **Perseus** to her rescue. To the south, beyond the intervening constellations of Aries and Pisces, Cetus, the sea-monster, who was to have devoured her, stretches his ungainly bulk.

Names of Stars. α, *Alpheratz;* β, *Mirach;* γ, *Almaach.*

Double Stars. (1) γ, Mags. 3, 5; Pos. 62°; Dist. 11″; colors, orange and greenish blue—a beautiful object. The small star is itself double, but at present so close as to be beyond the reach of any but very large instruments (Astr. Fig. 113). (2) π (2° N. and a little west of δ), Mags. 4, 9; Pos. 174°; Dist. 36″; white and blue.

Nebulæ. M. 31; the great nebula; visible to naked eye. M. 32; small, round, and bright, is in the same low-power field with 31; south and east of it.

17. Pisces (Map II.).—Immediately south of Andromeda lies Pisces, the first of the zodiacal constellations, though now occupying (in consequence of precession) the *sign* of Aries. It has not a single conspicuous star, and is notable only as containing the vernal equinox, or first of Aries, which lies near the southern boundary of the constellation in a peculiarly starless region. A line from α Andromedæ through γ Pegasi continued as far again strikes about 2° east of the equinox.

The body of the southern fish lies about 15° south of the middle of the southern side of the great square of Pegasus, and is marked by an irregular polygon of small stars, 5° or 6° in diameter. A long crooked "ribbon" of little stars runs eastward for more than 30°, terminating in α Piscium, called *El Rischa,* a star of the fourth magnitude 20° south of the head of Aries. From there another line of stars leads up N.W. in the direction of δ Andromedæ to the northern fish, which lies in the vacant space south of β Andromedæ.

Double Stars. (1) α, Mags. 4, 5.5; Pos. 324°; Dist. 3″. (2) ψ′ (2° S.E. of η Andromedæ—see map), Mags. 4.9, 5; Pos. 160°; Dist. 31″.

18. Triangulum (Map II.).—This little constellation, insignificant as it is, is one of Ptolemy's ancient 48. It lies half-way between γ Andromedæ and the head of Aries, characterized by three stars of the third and fourth magnitudes.

Double Stars. (1) ι or 6 (5° nearly due south of β Trianguli, and
at the obtuse angle of an isosceles triangle of which α and γ are the
other two corners), Mags. 5, 6.5; Pos. 76°; Dist. 4″; topaz-yellow
and green.

19. Aries (Map II.). — This is the second of the zodiacal
constellations (now occupying the *sign* of Taurus). It is bounded
north by Triangulum and Perseus, west by Pisces, south by
Cetus, and east by Taurus. The characteristic star-group is
that composed of α, β, γ (see map), about 20° due south of γ
Andromedæ: α, a star of the 2½ magnitude is fairly conspicuous, forming as it does a large isosceles triangle with β and γ
Andromedæ.

Names of Stars. α, *Hamal*; β, *Sheratan*; γ, *Mesartim.*
Double Stars. (1) γ, Mags. 4.5, 5; Pos. 0°; Dist. 8″.8. (This is
probably the earliest known double star; noticed by Hooke in 1664.)
(2) ε, Mags. 5, 6.5; Pos. 200°; Dist. 1″.2. (About one-third of the
way from α Arietis towards Aldebaran, ζ is 4° beyond it on the same
line.) This is probably too difficult for any instrument less than 4 or
4½ inches' aperture. (3) π, Triple; Mags. 5, 8.5, and 11; A–B, Pos.
122°; Dist. 3″.1; A–C, Pos. 110°; Dist. 25″. (At the southern corner
of a nearly isosceles triangle formed with ε and τ, ε being at the obtuse
angle.)

The star 41 Arietis (3½ mag.), which forms a nearly equilateral triangle with α Arietis and γ Trianguli, constitutes, with two or three
other small stars near it, the constellation of *Musca* (Borealis), a constellation, however, not now generally recognized.

20. Cetus (Map II.). — South of Aries and Pisces lies the
huge constellation of Cetus, which backs up into the sky from
the southeastern horizon. The head lies some 20° S.E. of α
Arietis, marked by an irregular pentagon of stars, each side of
which is 5° or 6° long. The southern edge of it is formed by
the stars α (2½ mag.) and γ (3½ mag.) : δ lies nearly south of
γ. β, the brightest star of the constellation (2d magnitude),
stands alone nearly 40° west and south of α. About half-way

from β to γ the line joining them passes through a characteristic quadrilateral (see map), the N.E. corner of which is composed of two fourth-magnitude stars, ζ and χ. The remarkable variable o Ceti (Mira) lies almost exactly on the line joining γ and ζ, a little nearer to γ than to ζ. It is visible to the naked eye for about a month or six weeks every eleven months, when near its maximum.

Names of Stars. a, MENKAR; β, *Diphda* or *Deneb Kaitos;* ζ, *Baten Kaitos;* o, MIRA.

Double Stars. (1) γ, Mags. 3.5, 7; Pos. 290°; Dist. 2″.5; yellow and blue.

South of Cetus lies the constellation of Sculptoris Apparatus (usually known simply as Sculptor), which, however, contains nothing that requires notice here. South of Sculptor, and close to the horizon, even when on the meridian, is Phœnix. It has some bright stars, but none easily observable in the United States.

21. Perseus (Maps I. and II.). — Returning now to the northern limit of the map, we come to the constellation of Perseus. Its principal star is a, rather brighter than the standard second magnitude, situated very nearly on the prolongation of the line of the three chief stars of Andromeda. A very characteristic configuration is "the segment of Perseus" (Map I.), a curved line, formed by δ, a, γ, and η, with some smaller stars, concave towards the northeast, and running along the line of the Milky Way towards Cassiopeia. The remarkable variable star β, or Algol (Astr. 453), is situated about 9° south and a little west of a, at the right angle of a right-angled triangle which it forms with a (Persei) and γ Andromedæ. Some 8° south and slightly east of δ is ϵ, and 8° south of ϵ are ζ and o of the fourth magnitude in the foot of the hero. Algol and a few small stars near it form "Medusa's Head."

Names of Stars. a is *Marfak*, or ALGENIB; β is ALGOL.

Double Stars. (1) ϵ, Mags. 3.5, 9; Pos. 10°; Dist. 8″.4. (2) ζ,

Quadruple; Mags. 3.5, 10, 11, 12; Pos. A–B, 207°; Dist. 13″.2, 83″, 121″. (3) η, Mags. 5, 8.5; Pos. 300°; Dist. 28″; orange and blue.

Clusters. (1) ⚳ VI. 33 and 34. Magnificent. Half-way between γ Persei and δ Cassiopeiæ. (2) M. 34; A.R. 2ʰ 34ᵐ; Dec. 42° 11′; coarse, with a pretty double star (eighth mag.) included.

22. Aurīga (Maps I. and II.). — Proceeding east from Perseus we come to Auriga, instantly recognized by the bright yellow star CAPELLA (the Goat) and her attendant Hœdi (or Kids). Capella, a Aurigæ, according to Pickering, is precisely of the same brightness as Vega (Mag. = 0.2), both of them being about ⅓ of a magnitude fainter than Arcturus, but distinctly brighter than any other stars visible in our latitudes except Sirius itself. About 10° east of Capella is β Aurigæ of the second magnitude, and 8° south of β is θ of the third magnitude; δ Aurigæ is 10° north of β in the circumpolar region. ε, ζ, and η, 4° or 5° S.W. of a, are the "Kids."

Names of Stars. a, CAPELLA; β, MENKALINAN.

Double Stars. (1) ω, Mags. 5, 9; Pos. 353°; Dist. 7″; white, light blue. β is a spectroscopic double (see Art. 465*).

Clusters. (1) M. 37; A.R. 5ʰ 44ᵐ; Dec. 32° 31′ (on the line from θ Aurigæ to ζ Tauri, one-third of the way from θ). Fine for small instrument. (2) M. 38; A.R. 5ʰ 21ᵐ; Dec. 35° 47′. Nearly at the middle of the line from θ to ω. (3) M. 36; A.R. 5ʰ 28ᵐ; Dec. 34° 3′. One-third of the way from M. 38 to M. 37.

23. Taurus (Map II.). — This, the third of the zodiacal constellations, is bounded north by Perseus and Auriga, west by Aries, south by Eridanus and Orion, and east by Orion and Gemini. It is unmistakably characterized by the Pleiades, and by the V-shaped group of the Hyades which forms the face of the bull, with the red ALDEBARAN (a Tauri) blazing in the creature's eye, as he charges down upon Orion. His horns reach out towards Gemini and Auriga, and are tipped with the second and third magnitude stars β and ζ. As in the case of Pegasus, only the head and shoulders appear in the constellation. Six of the Pleiades are easily visible, and on a dark night a fairly good eye will count nine (see Astr. 469). With a 3-inch telescope about 100 stars are visible in the cluster. In the Hyades the pretty naked-eye double θ_1, θ_2, is worth noting.

Names of Stars. α, ALDEBARAN; β, *El Nath;* η (the brightest of the Pleiades), ALCYONE. For the names of the other Pleiades, see the figure in Art. 469 of the Astronomy.

Double Stars. (1) α has a small, distant companion, 12th magnitude; Pos. 36°; Dist. 1′ 48″. It has also a second companion much nearer and more minute, but far beyond the reach of ordinary telescopes. (2) τ, Mags. 5 and 8; Pos. 210°; Dist. 62 ′; white and violet. Found by drawing a line from γ (at the point of the V of the Hyades) through ε, and producing it as far again.

Nebula. M. 1; A.R. 5ʰ 27ᵐ; Dec. 21° 56′, about 1° west and a little north of ζ. Often mistaken for a comet. The so-called "Crab Nebula."

24. Orion (not O′-rĭ-on) (Map II.). — On the whole this is the finest constellation in the heavens. As he stands facing the bull his shoulders are marked by the two bright stars, α and γ, the former of which in color and brightness closely matches Aldebaran. In his left hand he holds up the lion skin, indicated by the curved line of little stars between γ and the Hyades. The top of the club, which he brandishes in his right hand, lies between ζ Tauri and μ and η Geminorum. His head is marked by a little triangle of stars of which λ is the chief. His belt consists of three stars of the second magnitude pointing obliquely downward towards Sirius. It is very nearly 3° in length, with the stars in it equidistant like a measuring-rod, so that it is known in England as the "Ell and Yard." From the belt hangs the sword, composed of three smaller stars lying more nearly north and south : the middle one of them is the multiple θ in the great nebula. β Orionis, or RIGEL, a magnificent white star, is in the left foot, and κ is in the right knee. Orion has no right *foot*, or if he has, it is hidden behind Lepus. The quadrilateral α, γ, δ, κ, with the diagonal belt δ, ε, ζ, once learned can never be mistaken for anything else in the heavens.

25. *Names of Stars.* α, BETELGEUSE; β, RIGEL; γ, *Bellatrix*; κ, *Saiph;* δ, *Mintaka;* ε, *Alnilam;* ζ, *Alnitak.*

Double Stars. In these Orion is remarkably rich. (1) β (Rigel), Mags. 1 and 9; Pos. 200°; Dist. 9″.5; both white,—a beautiful and easy object. (2) δ (the westernmost star in the belt), Mags. 2.5 and 7; Pos. 0; Dist. 53″. (3) ζ, Triple; Mags. 2.5, 6.5, 10; A–B, Pos. 155°, Dist. 2″.4; A–C, Pos. 9°, Dist. 59″. (4) ι, Triple; Mags. 3.5, 8.5, 11; A–B, Pos. 142°, Dist. 11″.5; A–C, Pos. 103°, Dist. 49″. (This is the lowest star in the sword, just below the nebula.) (5) θ, Multiple, the trapezium in the nebula. Four stars are easily seen by small telescopes (Astr. Fig. 113). (6) σ, Triple; Mags. 4, 8, 7; A–B, Pos. 84°, Dist. 12″.5; A–C, Pos. 61°, Dist. 42″ (1¼° S.W. of ζ).

Nebula. M. 42; attached to the multiple star θ. *The* nebula of all the heavens; by far the finest known, though in a small telescope wanting much of the beauty brought out by a larger one.

26. Eridănus (Map II.). —This constellation lies south of Taurus, in the space between Cetus and Orion, and extends far below the southern horizon. Its brightest star α (ACHERNAR) is never visible in the United States.

Starting with β of the third magnitude, about 3° north and a little west of Rigel (β Orionis), one can follow a sinuous line of stars, some of them of the third and fourth magnitudes, westward about 30° to the paws of Cetus, 10° south of α Ceti; there the stream turns at right angles southwards for 10°, then southeast for about 20°, and finally southwestward to the horizon. One could succeed in fully tracing it out, however, only by help of a map on a larger scale than the one we are able to present.

Names of Stars. β, *Cursa;* γ, *Zaurack.*

Double Stars. (1) 32; A.R. 3ʰ 48ᵐ, Dec. S. 3° 19′; Mags. 5, 7; Pos. 347°; Dist. 6″.6; yellow, blue; very fine. (2) o², Triple; A.R.

4^h 10^m; Dec. S. 7° $50'$; Mags. 5, 10, 10; Pos. A, $\dfrac{(B+C)}{2}$, 108°; Dist. $83''$; Pos. B–C, 110°, Dist. $4''$; very pretty.

27. Lepus (Map II.).—This little constellation (one of the ancient 48) lies just south of Orion, occupying a space of some 15° square. Its characteristic configuration is a quadrilateral of third and fourth magnitude stars, with sides from 3° to 5° long, 10° south of κ Orionis, and 15° west and a little south of Sirius.

Double Stars. (1) γ (the S.E. corner of the quadrilateral) is a coarse double. Mags. 4, 6.5; Pos. 350°; Dist. $93''$. (2) κ ($5\frac{1}{2}^\circ$ south of Rigel), Mags. 5 and 9; Pos. 0°; Dist. $3''.7$.

28. Columba (Noah's Dove) (Map II.).—This is next south of Lepus : too far south to be well seen in the Northern States. Its principal star, a, or *Phact*, of the $2\frac{1}{2}$ magnitude, is readily found by drawing a line from Procyon to Sirius, and prolonging it nearly the same distance. And in passing we may note that a similar line drawn from a Orionis through Sirius and produced, will strike near ζ Argus, or " *Naos*," a star about as bright as Phact,—the two lines which intersect at Sirius making the so-called "*Egyptian X.*"

29. Lynx (Map I., II., and III.).—Returning now to the northern limit of the map, we find the modern constellation of the Lynx lying just east of Auriga and enveloping it on the north and in the circumpolar region. It contains no stars above the fourth magnitude, and is of no importance except as occupying an otherwise vacant space.

Double Stars. (1) 38, or ρ Lyncis, A.R. 9^h 11^m; Dec. 37° $21'$; Mags. 4, 7.5; Pos. 240°; Dist. $2''.9$; white and lilac. (This is the northern one of a pair of stars which closely resembles the three pairs that mark the paws of Ursa Major. This pair makes nearly an isosceles triangle with the two pairs λ μ and ι κ, Ursæ Majoris — see map.)

30. Gemini (Map II.).—This is the fourth of the zodiacal constellations (mostly in the *sign* of Cancer), containing the summer solstitial point about 2° west and a little north of the star η. It lies northeast of Orion and southeast of Auriga, and is sufficiently characterized by the two stars a and β (about $4\frac{1}{2}^\circ$ apart), which mark the heads of the twins. The southern

one, β, or POLLUX, is the brighter, but a (CASTOR) is much the
more interesting, as being double. The feet are marked by
the third-magnitude stars γ and μ, some 10° east of ζ Tauri,
and the map shows how the lines that join these to β and a
respectively mark the places of δ and ϵ. η, 2° west of μ, is a
variable, and is also double, though as such it is beyond the
power of ordinary telescopes.

Names. a, CASTOR; β, POLLUX; γ, *Alhena*, μ, *Tejat (Post)*;
η, *Tejat (Prior)*; δ, *Wasat;* ϵ, *Meboula.*

Double Stars. (1) a, Mags. 2.5, 3; Pos. 225°; Dist. 5″.5. Binary;
period undetermined, but certainly over 200 years. The larger of
the close pair is also a spectroscopic binary, with period of about 3
days (see Art. 465*). There is also a companion of ninth mag., distant about 74″; Pos. 164°. (2) δ, Mags. 3, 8 ; Pos. 203°; Dist. 7″.
(3) μ, Mags. 3, 11 ; Pos. 79°; Dist. 80″.

Nebulæ and Clusters. (1) M. 35; A.R. 6ʰ 01ᵐ; Dec. 24° 21′; N.W.
of η at the same distance as that from μ to η, and on the line from γ
through η produced. The map is not quite right in this respect.
(2) ⚥ IV. 45; A.R. 7ʰ 22ᵐ; N. 21° 10′. A nebulous star in a small
telescope ; in a large telescope, very peculiar — 2° southeast of δ.

31. Canis Minor (Map II.). — This constellation, just south
of Gemini, is sufficiently characterized by the bright star
PROCYON, which is 25° due south of the mid-point between
Castor and Pollux. a, β, and γ together form a configuration
closely resembling that formed by a, β, and γ Arietis. Procyon, a Orionis, and Sirius form nearly an equilateral triangle
with sides of about 25°.

Names. a, PROCYON; β, *Gomelza.*
Double Stars. (1) Procyon has a small companion, Dist. 40″, Pos.
312°, — too small, however, for anything less than an 8-inch telescope.
In 1896 a still smaller companion, like that of Sirius, was found
much nearer the large star. (See Art. 464.) (2) (Σ 1126) (following
Procyon 43ˢ, and 2′ south, — the brightest of the stars in that field),
Mags. 7, 7.5 ; Pos. 145°; Dist. 1″.5 ; a good test for a 4-inch glass.

32. Monoceros (The Unicorn) (Map II.). — This is one of the new constellations organized by Hevelius to fill the gap between Gemini and Canis Minor on the north, and Argo Navis and Canis Major on the south. It lies just east of Orion. It has no conspicuous stars, but is traversed by a brilliant portion of the Milky Way. The α (fourth mag.) of the constellation lies about half-way between α Orionis and Sirius, a little west of the line joining them.

Double Stars. (1) 8, or *b* (7½° east and 3° south of α Orionis), Mags. 5, 8; Pos. 24°; Dist. 12″.9; colors, orange and lilac. A fine low-power field. (2) 11 Monocerötis, a fine *triple* (see Fig. 113 of Astr.), A.R. 6ʰ 24ᵐ; Dec. *south* 6° 57′; A to B-C, Pos. 130°, Dist. 8″; B-C, Pos. 120°, Dist. 2″.5. The star is very nearly pointed at by a line drawn from ζ Canis Majoris, north through β, and continued as far again.

Clusters. (1) ♅ VII. 2; A.R. 6ʰ 24ᵐ; Dec. N. 5° 2′ (visible to the naked eye about 1¼° N.E. of 8 Monocerotis described above). A fine cluster for a low power. (2) M. 50; A.R. 6ʰ 57ᵐ; Dec. S. 8° 9′. In the Milky Way, on the line from Sirius to Procyon, two-fifths of the distance.

33. Canis Major (Map II.). — This glorious constellation needs no description. Its α is the Dog Star, SIRIUS, beyond all comparison the brightest in the heavens, and probably one of our nearer neighbors. It is nearly pointed at by the line drawn through the three stars of Orion's belt. β, at the extremity of the uplifted paw, is of the second magnitude, and so are several of those farther south in the rump and tail of the animal, who sits up watching his master Orion, but with an eye out for Lepus.

Names. α, SIRIUS; β, *Mirzam;* γ, *Muliphen;* δ, *Wesen;* ε, *Adara.* γ is said to have disappeared from 1670 to 1690, but at present is not recognized as variable, though much fainter than would be expected from its being lettered as γ.

Double Stars. (1) Sirius itself has a small companion (see Art. 463). (2) μ (4° N.E. of Sirius), Mags. 5, 9.5 ; Pos. 335°; Dist. 3″.5.

Clusters. (1) M. 41 (4° south of Sirius) ; a fine group with a red star near centre.

34. Argo Navis (The Ship) (Maps II. and III.). — This is one of the largest, most important, and oldest of the constellations, lying south and east of Canis Major. Many Uranographers now divide it into *three*, Puppis, Vela, and Carina. Its brightest star, α Argûs, Canopus, ranks next to Sirius, but is not visible anywhere north of the parallel of 38°. The constellation, huge as it is, is only a *half* one, like Pegasus and Taurus, — only the stern of a vessel, with mast, sail, and oars; the stem being wanting. In the part of the constellation covered by our maps the most conspicuous stars lie east and southeast of Canis Major. We have already mentioned ζ, or *Naos* (Art. 28), at the southeast extremity of the "Egyptian X"; and about 8° south and a little east of it is γ, nearly of the second magnitude.

Clusters. One or two clusters are accessible in our latitudes. (1) (⚹ VIII. 38), A.R. $7^h 31^m$; Dec. S. 14° 12'. Pointed at by the line from β Can. Maj. through Sirius, continued 2½ times as far. Visible to naked eye; rather coarse. (2) M. 46; a little more than 1° east and south of the preceding. (3) M. 93, A.R. $7^h 39^m$; Dec. S. 23° 34'; about 2° N.W. of ζ Argûs.

35. Cancer (Maps II. and III.). — This is the fifth of the zodiacal constellations, bounded north by Lynx and Leo Minor, south by the head of Hydra, west by Gemini and Canis Minor, and east by Leo. It does not contain a single conspicuous star, but is easily recognizable from its position, and in a dark night by the nebulous cloud known as "Præsepe," or the "Manger," with the two stars γ and δ near it, — the so-called "Aselli," or "Donkeys." Præsepe (sometimes also called the "Beehive") is really a coarse cluster of seventh and eighth magnitude stars, resolvable by an opera-glass. The line from Castor through Pollux, produced about 12°, passes near enough to it to serve as a pointer. α, of the fourth magnitude, is on the line drawn from Præsepe through δ (the southern Asellus), produced about 7°; β may be recognized by drawing a line from γ (the northern Asellus) through Præsepe, and continuing it about 12°.

Double Stars. (1) ι, Mags. 4, 6.5; Pos. 308°; Dist. 30'; orange and blue; nearly due north of γ, distance twice that between the Aselli.

(2) ζ, Triple (see Astr. Fig. 113); A–B, Mags. 6 and 7, Pos. (1905) 346°, Dist. 1″; in rapid motion; period about 60 years. A–C, Pos. (1905) 110°, Dist. 5″; also in motion, but period unknown and much longer. Easily found by a line from α Gem. through β, produced two and a half times as far.

36. Leo (Map III.).—East of Cancer lies the noble con-stellation of Leo, which adorns the evening sky in March and April; it is the sixth of the zodiacal constellations, now occu-pying the sign of Virgo. Its leading star REGULUS, or "*Cor Leonis*," is of the first magnitude, and two others, β and γ, are of the second. α, γ, δ, and β form a conspicuous irregular quadrilateral (see map), the line from Regulus to Denebola being 26° long. Another characteristic configuration is "The Sickle," of which α, η is the handle, and the curved line η, γ, ζ, μ, and ε is the blade, the cutting edge being turned towards Cancer. The "radiant" of the November meteors lies between ζ and ε.

Names. α, REGULUS; β, DENEBOLA; γ, *Algeiba;* δ, *Zosma.*
Double Stars. (1) γ, Mags. 2, 3.5; Pos. 116°; Dist. 3″.4; binary; period about 400 years. (2) ι, Mags. 4 and 7; Pos. 65°; Dist. 2″.5; yellow and bluish; easily recognized by aid of the map. (3) 54, Mags. 4.5, 7; Pos. 103°; Dist. 6″.2. Found by producing the line from β through δ half its length.

37. Leo Minor and Sextans (Map III.).—*Leo Minor* is an insignificant modern constellation composed of a few small stars north of Leo, between it and the hind feet of Ursa Major. It contains nothing deserving special notice. A similar remark holds as to *Sextans* even more emphatically.

38. Hydra (Map III.).—This constellation, with its riders Crater and Corvus, is a large and important one, though not very brilliant. The head is marked by a group of five or six fourth and fifth magnitude stars just 15° south of Præsepe. A curving line of small stars leads down southeast to α, "*Cor Hydræ*," a small second or bright third magnitude star stand-

ing very much alone. From there, as the map shows, an
irregular line of fourth-magnitude stars running far south and
then east, almost to the boundary of Scorpio, marks the creature's body and tail, the whole covering almost six hours of
right ascension, and very nearly 90° of the sky. About the
middle of his length, and just below the hind feet of Leo (30°
due south from Denebola), we find the little constellation of
Crater; and just east of it the still smaller but much more
conspicuous one of *Corvus*, with two second-magnitude stars
in it, and four of the third and fourth magnitudes. It is well
marked by a characteristic quadrilateral (see map); with δ and
η together at its northeast corner. The order of the letters
differs widely from that of brightness in this constellation,
suggesting that changes may have occurred.

Names. a Hydræ, ALPHARD or *Cor Hydræ;* a Crateris, *Alkes;*
a Corvi, *Alchiba;* δ Corvi, *Algores.*

Double Stars. (1) ε Hydræ (the northernmost one of the group
that marks the head), Mags. 4, 8; Pos. 220°; Dist. 3″.5; yellow and
purple. (2) δ Corvi, Mags. 3, 8; Pos. 210°; Dist. 24″; yellow
and purple. (3) *Nebula,* ♉ IV. 27, A.R. 10ʰ 19ᵐ; Dec. S. 18° 2′ (3°
S. and ¼° W. of μ — see map). Bright planetary nebula, about as
large as Jupiter.

39. Virgo (Map III.). — East and south of Leo lies Virgo,
the seventh zodiacal constellation, bounded on the north by
Boötes and Coma Berenicis, on the east by Boötes and Libra,
and on the south by Corvus and Hydra. Its a, SPICA Virginis, is of the 1½ magnitude and, standing rather alone 10°
south of the celestial equator, is easily recognized as the
southern apex of a nearly equilateral triangle which it forms
with Denebola (β Leonis) to the northwest, and Arcturus
northeast of it. β Virginis of the third magnitude is 14° due
south of Denebola. A line drawn eastward and a little south
from β (third magnitude) and then carried on, curving north-
ward, passes successively (see map) through η, γ, δ, and ε,

of the third magnitude (notice the word formed by the letters *Bĕgde*, like *Bagdei* in Cassiopeia, Art. 9). θ lies nearly midway between α and δ. There are also a number of other fourth-magnitude stars.

Names. α, Spica and *Azimech; β, Zavijava; ε, Vindemiatrix.*

Double Stars. (1) γ, (Binary; period 185 years; not quite half-way from Spica to Denebola, and a little west of the line), Mags. 3, 3; Pos. (1905) 328°; Dist. 6″.2; very easy and fine (Astr. Fig. 113). (2) θ (two-fifths of the way from *Spica* towards δ), Triple; Mags. A 4.5, B 9, C 10; Pos. A–B, 345°, Dist. 7″; A–C, Pos. 295°, Dist. 65″. (3) o (one-third of the way from *Denebola* towards γ Virginis), Mags. 6 and 8; Pos. 228°; Dist. 3″.5. Spica is a spectroscopic binary (Art. 465*).

Nebulæ. (1) M. 49; A.R. 12ʰ 24ᵐ; Dec. + 8° 40′. Forms an equilateral triangle with δ and ε. It lies in the remarkable "nebulous" region of Virgo. But most of the nebulæ are faint, and observable only with large telescopes. (2) ♅ II. 74 and 75; A.R. 12ʰ 47ᵐ; Dec. + 11° 53′; two in one field, 2° west and a little north of ε. (3) M. 86 (midway between Denebola and ε); A.R. 12ʰ 20ᵐ; Dec. + 13° 36′. A large telescope shows nearly a dozen nebulæ within 2° of this place.

40. Coma Berenicis (Map III.). — This little constellation, composed of a great number of fifth and sixth magnitude stars, lies 30° north of γ and η Virginis, and about 15° northeast of Denebola. It contains a number of interesting double stars, but they are not easily found without the help of an equatorial mounting and graduated circles.

41. Canes Venatici (The Hunting Dogs). — These are the dogs with which Boötes is pursuing the Great Bear around the pole: the northern of the two is *Asterion*, the southern *Chara*. Most of the stars are small, but α is of the 2½ magnitude, and is easily found by drawing from η Ursæ Majoris (the star in the end of the Dipper-handle) a line to the southwest, perpendicular to the line from η to ζ (Mizar) and about 15° long: it is about one-third of the way from η Ursæ Majoris to δ Leonis. With Arcturus and Denebola it forms a triangle much like that which they form with Spica.

Names. a is known as Cor Caroli (Charles II. of England).

Double Stars. (1) a, or 12 Canum, Mags. 3 and 5; Pos. 227°; Dist. 20″. (2) 2 Canum (one-third of the way from a towards δ Leonis), Mags. 6 and 8; Pos. 260°; Dist. 41″.3; orange, smalt blue.

Nebulæ. (1) M. 51; A.R. 13ʰ 25ᵐ; Dec. 47° 49′ (3° west and somewhat south of Benetnasch). A faint double nebula in small telescopes; in great ones, the wonderful "Whirlpool Nebula" of Lord Rosse. (2) M. 3; bright cluster (half a degree north of the line from a Canum to Arcturus, and a little nearer the latter). It is one of the *variable-star clusters* discovered in 1895 (see Art. 455*).

42. Boötes (Maps III. and I.). — This fine constellation is bounded on the west by Ursa Major, Canes Venatici, Coma Berenicis, and Virgo, and on the south by Virgo. It extends more than 60° in declination, from near the equator quite to Draco, where the uplifted hand overlaps the tail of the Bear. Its principal star, ARCTURUS, is of a ruddy hue, and in brightness is excelled only by Sirius among the stars visible in our latitudes. Canopus and a Centauri are reckoned brighter, but they are southern circumpolars. Arcturus is at once recognized by its forming with Spica and Denebola the great triangle already mentioned (Art. 39). Six degrees west and a little south of it is η, of the third magnitude, which forms with it, in connection with ʋ, a configuration like that in the head of Aries. ε is about 10° northeast of Arcturus, and in the same direction about 10° farther lies δ. A pentagon is formed by these two stars along with β, γ, and ρ. "Boötes" means "the shouter" (or, according to others, "the herdsman").

Names. a, ARCTURUS; β, *Nekkar;* ε, *Izar;* η, *Muphrid;* γ, *Seginus.*

Double Stars. (1) ε, Mags. 3, 6; Pos. 325°; Dist. 3″.1; orange and greenish blue; very fine. (2) ζ (about 9° southeast from Arcturus, at right angles to the line aε), Mags. 3.5, 4; Pos. 295°; Dist. 0″.8; a good test for a 4-inch glass. (3) π (2½° north of ζ), Mags. 4.9, 6; Pos. 101°; Dist. 5″.3. (4) ξ (10° due east from Arcturus, 3° N.E. from π), Mags. 4.7, 6.6; Pos. (1905) 176°; Dist. 2″.5; yellow and purple. Binary; period 127 years.

43. Corona Borealis (Map III.). — This beautiful little constellation lies 20° northeast of Arcturus, and is at once recognizable as an almost perfect semicircle composed of half a dozen stars, among which the brightest, α, is of the second magnitude. The extreme northern one is θ; next comes β, and the rest follow in the β α γ δ ε ι (*Bagdei*) order, just as in Cassiopeia.

Names. α, *Gemma*, or *Alphacca*.
Double Stars. (1) ζ (nearly pointed at by ε–δ Boötis; 7° from ε), Mags. 5, 6; Pos. 301°; Dist. 6″; white and greenish. (2) η, rapid binary, at certain times can be split by a 4-inch glass. Mags. 6, 6.5; pointed at by the line from α through β, 2° beyond β. The temporary star of 1866 (Astr. 450) lies 1½° S.E. of ε Coronæ.

44. Libra (Map III.). — This is the eighth of the zodiacal constellations, and lies east of Virgo, bounded on the south by Centaurus and Lupus, on the east by the upstretched claw of Scorpio, and on the north by Serpens and Virgo. It is inconspicuous, the most characteristic figure being the trapezoid formed by the lines joining the four stars α, ι, γ, β. β, which is the northernmost of the four, is the brightest (2½ magnitude), and is about 30° nearly due east from Spica, while α is about 10° southwest of β. The remarkable variable δ Libræ is 4° west and a little north from β. Most of the time it is of the 4½ or fifth magnitude, but runs down nearly two magnitudes at the minimum.

Names. α, *Zuben el Genubi*; β, *Zuben el Chamali*.
Cluster. M. 5; A.R. 15ʰ 12ᵐ; Dec. N. 2° 32′. This is within the boundaries of *Serpens*, and just a little north and west of the fifth-magnitude star 5 Serpentis. It is a *variable-star cluster* (Art. 455*).

45. Antlia, Centaurus, and Lupus (Map III.). — These constellations lie south of Hydra and Libra. *Antlia Pneumatica* (the " Air-Pump ") is a modern constellation of no importance and hardly recognizable by the eye, having only a single star as bright as the 4½ magnitude. *Centaurus*, on the other hand, is an ancient and extensive asterism, containing in its (south) circumpolar portion two stars of the first magnitude : α Centauri stands next after Sirius and Canopus in brightness, and, as far as present knowledge indicates, is our nearest neighbor among the stars. The part of the constellation which becomes visible in our latitudes is not specially brilliant, though it contains several stars of the 2½ and third magnitude in the region that lies south of Corvus and Spica Virginis. A line from ε Virginis through Spica, produced a little more than its own length, will strike very near θ, a solitary star of the 2½ magnitude in the Centaur's left shoulder. ι (third mag.) lies 11° west of θ, and η (third mag.) 9° southeast; while 5° or 6° south of the line from θ to ι lies a tangle of third-magnitude stars, which, if they were at a higher elevation, would be conspicuous. Centaurus is best seen in May or early in June.

Lupus, also one of Ptolemy's constellations, lies due east of Centaurus and just south of Libra. It contains a considerable number of third and fourth magnitude stars ; but is too low for any satisfactory study in our own latitudes. It is best seen late in June. These constellations contain numerous objects interesting for a southern observer, but nothing available for our purpose.

46. Scorpio (*or Scorpius*) (Map III.). — This, the ninth of the zodiacal constellations, and the most brilliant of them, lies southeast of Libra, which in ancient times used to form its claws (*Chelœ*). It is bounded north by Ophiuchus, south by Lupus, Norma, and Ara, and east by Sagittarius. It is recognizable at once on a summer evening by the peculiar configuration, like a boy's kite, with a long streaming tail reaching far down to the southern horizon. Its principal star, ANTARES, is of the first magnitude and fiery red, like the planet Mars. From this it gets its name, which means "the rival of *Ares*" (*Mars*). β (second magnitude) is in the arch of the kite bow, about 8° or 9° northwest of Antares, while the star which Bayer lettered as γ Scorpii is well within Libra, 20° west of Antares. (There is no little discordance and confusion among Uranographers as to the boundary between the two constellations.) The other principal stars of the constellation are easily found on the map; δ is 3° southwest of β, while ε, ζ, η, θ, ι, κ, and λ follow along in order in the tail of the creature, except that between ε and ζ is interposed the double μ. ε, θ, and λ are all of the second magnitude, and the others of the third.

47. *Names.* a, ANTARES; β, *Akrab.*

Double Stars. (1) a, Mags. 1 and 7; Pos. 270°; Dist. 3."5; fiery red and vivid green. A beautiful object when the state of the air allows it to be well seen. (2) β, Triple; Mags. A 2, B 4, C 10; A–B, Pos. 25°, Dist. 13''; A–C, Pos. 89°, Dist. 0''.9. (3) ν (2° due east of β), Quadruple; Mags. A 4, B 5, C 7, D 8; A–B, Pos. 7°, Dist. 0''.8; A–C, Pos. 337°, Dist. 41''; C–D, Pos. 47°, Dist. 2''.4. A beautiful object. (4) ξ Scorpii (8½° due north from β), Triple; Mags. A 5, B 5.2, C 7.5 ; A–B, (Binary) Pos. 200°, Dist. 1''.4 ; ½(A + B) to C; Pos. 65°, Dist. 7''.3. μ¹ is a spectroscopic binary (Art. 465*).

Clusters. (1) M. 80, A.R. 16ʰ 10ᵐ; Dec. S. 22° 42'; half-way between a and β; one of the finest clusters known. (2) M. 4, A.R. 16ʰ 16ᵐ; Dec. S. 26° 14'; 1½° west of a; not so fine as the preceding.

Norma lies west of Scorpio, between it and *Lupus*, while *Ara* lies due south of η and θ. Both are small and of little importance, at least to observers in our latitudes.

48. Ophiūchus (*or Serpentarius*) and **Serpens** (Map III.). — Ophiuchus means the "serpent-holder." The giant is represented as standing with his feet on Scorpio, and grasping the "serpent," the head of which is just south of Corona Borealis, while the tail extends nearly to Aquila. The two constellations therefore are best treated together. The head of Serpens is marked by a group of small stars 20° due east of Arcturus, and 10° south of Corona. β and γ are the two brightest stars in the group, their magnitudes three and a half and four. δ lies 6° southwest of β, and there the serpent's body bends southeast through a and ε Serpentis (see map) to δ and ε Ophiuchi in the giant's hand. The line of these five stars carried upwards passes nearly through ε Boötis, and downwards through ζ Ophiuchi.   A line crossing this at right angles, nearly midway between ε Serpentis and δ Ophiuchi, passes through μ Serpentis on the southwest, and λ Ophiuchi to the northeast.   The lozenge-shaped figure formed by the lines drawn from α Serpentis and ζ Ophiuchi to the two stars last mentioned forms one of the most characteristic configurations of the summer sky.

α Ophiuchi (2½ magnitude) is easily recognized in connection with α Herculis, since they stand rather isolated, about 6° apart, on the line drawn from Arcturus through the head of Serpens, and produced as far again.   α Ophiuchi is the eastern and the brighter of the two.   It forms with Vega and Altair a nearly equilateral triangle.   β Ophiuchi lies about 9° southeast of α; and 5° east and a little south of β are five small stars in the Milky Way, forming a V with the point to the south, much like the Hyades of Taurus.   They form the head of the now discredited constellation "Poniatowski's Bull" (*Taurus Poniatovii*), proposed in 1777.

49. *Names.*   α Ophiuchi, *Ras Alaghue;* β, *Cebalrai;* δ, *Yed;* λ, *Marfic:* α Serpentis, *Unukalhai;* θ, *Alya.*

Double Stars.   (1) λ Ophiuchi, Binary; period, 234 years; Mags. 4, 6; Pos. (1905) 60°; Dist. 1″.2.   (2) 70 Ophiuchi (the middle star in the eastern leg of the V of Poniatowski's Bull),   Binary; period, 93 years; Mags. 4.5, 7; Pos. (1905) 180°; Dist. 2″.   The position angle changes very rapidly just now, and the star is too close to be resolved by a small instrument.   (3) δ Serpentis, Mags. 4, 5; Pos. 185°; Dist. 3″.6; very pretty.   (4) θ Serpentis, Mags. 4, 4.5; Pos. 104°; Dist. 21″.   (5) ν Serpentis (4° N.E. of η Ophiuchi), Mags. 4.5, 9; Pos. 31°; Dist. 51″; sea-green and lilac.

Clusters.   (1) M. 23, A.R. 17ʰ 50ᵐ; Dec. S. 19° 0′.   Fine low-power field.   (2) M. 12, A.R. 16ʰ 41ᵐ; Dec. S. 1° 45′.   On the line between β and ε Ophiuchi, one-third of the way from ε.   (3) M. 10, A.R. 16ʰ 51ᵐ; Dec. S. 3° 56′.   On the line between β and ζ Ophiuchi, two-fifths of the way from ζ.   (4) ⚥ VIII. 72, A.R. 18ʰ 22ᵐ; Dec. N. 6° 29′.   Pointed at by the eastern leg of the Poniatowski V.   8° from 70 Ophiuchi.

50. Hercules (Maps I. and III.). — This noble constellation
lies next north of Ophiuchus, and is bounded on the west by
Serpens, Corona, and Boötes, while to the east lie Aquila, Lyra,
and Cygnus. On the north is Draco. The hero is represented
as resting on one knee, with his foot on the head of Draco,
while his head is close to that of Ophiuchus. The constellation contains no stars of the first or even of the second magnitude, but a number of the third. The most characteristic
figure is the keystone-shaped quadrilateral formed by the stars
ϵ, ζ, η, with π and ρ together at the northeast corner. It lies
about midway on the line from Vega to Corona. The line $\pi\epsilon$,
carried on 11°, brings us to β, the brightest star of the asterism; and γ and κ lie a few degrees farther along on the same
line continued toward γ Serpentis. The angle $\epsilon\beta\alpha$ is a right
angle opening towards Lyra. α is irregularly *variable*, besides
being *double*.

51. *Names.* α, *Ras Algethi*; β, *Korneforos*.

Double Stars. (1) α, Mags. 3, 6; Pos. 119°; Dist. 4″.5; orange and
blue. A very beautiful object for a 4-inch glass (Astr. Fig. 113). (2) ζ
(the S.W. corner of the "Keystone"), Binary; period, 34 y. (Astr. Fig.
113); Mags. 3, 6.5; Pos. (1905)190°; Dist. 1″.5. Rather difficult for a
small instrument. (3) ρ (2½° east of π at the N.W. corner of the
"Keystone"), Mags. 4, 5; Pos. 312°; Dist. 4″; white, emerald green.
(4) δ (on the line from η through ϵ produced nearly its own length),
Mags. 3, 8; Pos. 184°; Dist. 18″; white, light blue. Apparently an
"optical pair"; the relative motion being rectilinear. (5) μ (nearly
midway between Vega and α Herculis — see map), *Triple*; Mags. A 4,
B 9.5, C 10; A, $\frac{B+C}{2}$, Pos. 246°; Dist. 31″. B–C, too faint and close
for separation by any but large telescopes; Dist. about 1″; position
angle rapidly changing — about 20° in 1890. (6) 95 Herculis (the
N.W. corner of a little quadrilateral [sides 1° to 2°] of fourth and
fifth mag. stars, on line from ρ through μ, produced two-thirds its
length), Mags. 5.5 and 6; Pos. 262°; Dist. 6″; light green, cherry-red.
Peculiar in showing contrast of color between *nearly equal components*.

Clusters. (1) M. 13, A.R. 16ʰ 37ᵐ; Dec. 36° 41′. Exactly on the
western boundary of the Keystone, one-third the way from η towards

ζ. On the whole, the finest of all star clusters. (2) M. 92, A.R. 17ʰ 13ᵐ; Dec. 43° 16′ (6° north and a little west of ρ). Fine, but not equal to the other.

52. Lyra (Map IV.). — The great white or blue star VEGA sufficiently marks this constellation. It is attended on the east by two fourth-magnitude stars, ε and ζ, which form with it a little equilateral triangle having sides about 2° long. β and γ of the third magnitude (β is variable) lie about 8° southeast from Vega, 2½° apart.

Double Stars. (1) Vega itself has a small companion, 11th mag.; Pos. 160°; Dist. 48″. Only optically connected; the small star does not share the proper motion of the larger one, and has been used as a reference point in measuring Vega's parallax. (2) β, multiple; *i.e.*, it has three small stars near it, forming a very pretty object with a low power. (3) ε₁ and ε₂, Quadruple (the northern of the two which form the little triangle with α.) A sharp eye unaided by a telescope splits the star, and a small telescope divides both the components (see Astr. 468, and Fig. 113): ε₁ (or 4 Lyræ), Mags. 6, 7; Pos. 12°; Dist. 3″.2. ε₂ (or 5 Lyræ), Mags. 5.5, 6; Pos. 132°; Dist. 2″.5. ε₁ ε₂, Pos. 173°; Dist. 207″. On the whole, the finest object of the kind. (4) ζ, Mags. 4, 6; Pos. 150°; Dist. 44″. (5) η (10° E. of Vega), Mags. 4.5, 8; Pos. 85°; Dist. 28″; yellow, indigo. (6) δ; fine field for low powers.

Nebula. M. 57, the Annular Nebula. A.R. 18ʰ 49ᵐ; Dec. 32° 53′. Between β and γ, one-third of the way from β. (Art. 470.)

53. Cygnus (Maps I. and IV.). — This lies due east from Lyra, and is easily recognized by the cross that marks it. The bright star α (1½ magnitude) is at the top, and β (third magnitude) at the bottom, while γ is where the cross-bar from δ to ε intersects the main piece, which lies along the Milky Way from the northeast to the southwest. ζ is (nearly) on the prolongation of the line from γ through ε, not quite so far from ε as ε from γ.

Names. α, *Arided*, or *Deneb Cygni* (there are other *Denebs; e.g.*, *Deneb Kaitos* in Cetus); β, *Albireo;* γ, *Sadr.*

Double Stars. (1) β, Mags. 3.5, 7; Pos. 56°; Dist. 35″; orange,
smalt blue.   This is the finest of the colored pairs for a small telescope.   (2) μ (as far beyond ζ as ζ is east of ε, at the tip of the eastern wing), Mags. 5 and 6; Pos. 118°; Dist. 3″.8.   (3) χ (one-third of
the way from β towards γ), Mags. 5 and 9; Pos. 73°; Dist. 26″; yellow and blue.   (4) 61 Cygni (easily found by completing the parallel-
ogram of which α, γ, and ε are the other three corners.   σ and τ form
a little triangle with 61, which is the faintest of the three), Mags.
5.5, 6; Pos. (1905) 127°; Dist. 22″.   This is the star of which Bessel
measured the parallax in 1838 (Astr. 521), — apparently our *second
nearest* neighbor.

    δ is also a fine double, but too difficult for an instrument of less
than six inches' aperture.

    Clusters. (1) M. 39, A.R. 21ʰ 28ᵐ; Dec. 47° 54′ (about 3° north of ρ;
ρ itself (fourth mag.) being found by drawing a line from δ through
α, and carrying it an equal distance beyond.   (2) ⚳ VIII. 56, A.R.
20ʰ 19ᵐ; Dec. 40° 20′.   Beautiful group, ¼° north and a little east of
γ.   The bright spots in the Milky Way all through Cygnus afford
beautiful fields for a low power.

54. Vulpecula et Anser (Map IV.). — This little constellation is one of those originated by Hevelius, and has obtained
more general recognition among astronomers than most of his
creations.   It lies just south of Cygnus, and is bounded to the
south by Delphinus, Sagitta and Aquila.

    It has no conspicuous stars, but contains one very interesting tele-
scopic object, — the "Dumb-Bell Nebula," — M. 27, A.R. 19ʰ 54ᵐ;
Dec. 22° 23′.   On a line from γ Lyræ through β Cygni, produced as far
again, where this line intersects another drawn from α Aquilæ through
γ Sagittæ, 3½° north and half a degree east of the latter star.

55. Sagitta (Map IV.). — This little asterism, though very inconspicuous, is one of the old 48.   It lies south of Vulpecula, and the two
stars α and β, which mark the feather of the arrow, lie nearly midway
between β Cygni and Altair, while its point is marked by γ, 5° farther
east and north.

    Double Stars. (1) ζ (¾° N.W. of δ, the middle star in the shaft of
the arrow), Mags. 5.5, 9; Pos. 312°; Dist. 8″.6: the larger star is itself
close double, distance about ¼″, making an interesting triple system.

56. Aquila (Map IV.). — This constellation lies on the celestial equator, east of Ophiuchus and north of Sagittarius and Capricornus. It is bounded on the east by Aquarius and Delphinus, and on the north by Sagitta. Its characteristic configuration is that formed by ALTAIR (the standard first-magnitude star), with γ to the north and β to the south. It lies about 20° south of β Cygni, and forms a fine triangle with Vega and α Ophiuchi.

Double Star. (1) π Aquilæ (1½° N.E. of γ), Mags. 6 and 7; Pos. 120°; Dist. 1″.5. Good test for 3½-inch glass.

Cluster. M. 11, A.R. 18ʰ 45ᵐ; Dec. S. 6° 24′. A fine fan-shaped group of stars in the Milky Way. A line carried from Altair through δ Aquilæ (see map), and prolonged once and a half as far again, will find it about 4° S.W. of λ.

The southern part of the region allotted to Aquila on our maps has been assigned to *Antinoüs*. This constellation was recognized by some even in Ptolemy's time; but he declined to adopt it. Hevelius appropriated the eastern portion of "Antinoüs" for his constellation of "*Scutum Sobieski*," and M. 11 falls just within its limits.

57. Sagittarius (Map IV.). — This, the tenth of the zodiacal constellations, is bounded north by Aquila and Ophiuchus, west by Scorpio and Ophiuchus (though Bode and some other authorities crowd in a piece of "Telescopium" between it and Scorpio), south by Corona Australis, Telescopium, and Indus, and east by Microscopium and Capricornus. It contains no stars of the first magnitude, but a number of the 2½ and third.

The most characteristic configuration is the little inverted "milk dipper" formed by the five stars, λ, φ, σ, τ, and ζ, of which the last four form the bowl, while λ (in the Milky Way) is the handle. δ, γ, and ε, which form a triangle right-angled at δ, lie south and a little west of λ, the whole eight together forming a very striking group. There is a curious disregard of any apparent principle in the lettering of the stars of this constellation; α and β are stars not exceeding in brightness the fourth magnitude, about 4° apart on a north and south line and lying some 15° south and 5° east of ζ (see map). The Milky Way in Sagittarius is very bright, and complicated in structure, full of knots and streamers, and dark pockets.

Names. λ, *Kaus Borealis;* δ, *Kaus Media;* ε, *Kaus Australis;* σ, *Sádira.* This star is strongly suspected of irregular variability.

Double Stars. (1) μ¹ (7° N.W. of λ; on the line from ζ through φ produced), Triple; Mags. A 3.5, B 9.5, C 10; A–B, Pos. 315°, Dist. 40″; A–C, Pos. 114°, Dist. 45″.

Clusters and Nebulæ. (1) M. 22, A.R. 18ʰ 29ᵐ; Dec. S. 24° 0′ (3° N.W. of λ, and midway between μ and σ). Capital object for a 4-inch telescope. (2) M. 25, A.R. 18ʰ 25ᵐ; Dec. S. 19° 10′ (7° north and 1° east of λ; visible to naked eye). (3) M. 8; A.R. 17ʰ 57ᵐ; Dec. S. 24° 21′ (a little south of the line φλ produced, and as far from λ as λ from φ; also visible to naked eye). (4) ♅ IV. 41, The *Trifid Nebula,* A.R. 17ʰ 55ᵐ; Dec. S. 23° 2′ (1¼° north of M. 8, and almost exactly on the line φλ produced). A very beautiful and interesting object.

58. Capricornus (Map IV.). — This, the eleventh of the zodiacal constellations, follows Sagittarius on the east. It has Aquarius and Aquila (Antinoüs) on the north, Microscopium and Piscis Austrinus on the south, and Aquarius on the east. It has no bright stars, but the configuration formed by the two α's (α₁ and α₂) with each other and with β, 3° south, is characteristic and not easily mistaken for anything else. The two α's, a pretty "double" to the naked eye, lie on the line from β Cygni (at the foot of the cross) through Altair, produced about 25°. On the line αβ, about 3° distant, lies ρ (of the fourth magnitude), with two other small stars near it. From this a line 20° long, carried due east through θ and ι (of the fourth magnitude), brings the eye to γ and δ of the third, the latter marking the constellation's eastern limit.

Names. α, *Algiedi* (*prima* and *secunda*); δ, *Deneb Algiedi.*

Double Stars. (1) α₁ and α₂ (pretty with a very low power), Mags. 3 and 4; Dist. 6′ 13″. α₂ has also a very faint companion, invisible

with any telescope of less than 6-inch aperture; Pos. 150°; Dist. 7″.5.
The companion is itself double; Dist. about 1″; Pos. 240°.   (2) β,
Mags. 3.5, 7; Pos. 267°; Dist. 3′ 25″.   The companion is also a close
and difficult double.   (3) ρ (the northern star in the little triangle it
forms with π and o), Mags. 5, 9; Pos. 177°; Dist. 3″.8.   (4) π (the
S.W. one in the same triangle), Mags. 5, 9; Pos. 146°; Dist. 3″.5.

Nebula.   M. 30, A.R. 21ʰ 34ᵐ; Dec. S. 23° 42′ (about 1° west and
a little north of 41 Capricorni, a fifth-magnitude star, 7° south of γ
Capricorni).

59. Delphīnus (Map IV.). — This little asterism is ancient,
and unmistakably characterized by the rhombus of third-magnitude stars known as "Job's Coffin."   It lies about 15° northeast of Altair, bounded north by Vulpecula and west by Aquila.
There are a few stars visible to the naked eye in addition to
the four that form the rhombus.   *Epsilon*, about 3° southwest,
is the only conspicuous one.

Names.   a, *Svalocin;*   β, *Rotanev.*   These were given in joke by
Nicolaus Cacciatore, a Sicilian astronomer, about 1800.   The letters
of the two names *reversed* make Nicolavs Venator; Venator being the
translation of the Italian "Cacciatore," which means "*Hunter.*"   The
joke is good enough to keep.

Double Stars.   (1) γ (at the N.E. angle of the rhombus), Mags. 4,
7; Pos. 271°; Dist. 11″.3.   (2) β, a very close and rapid binary,
beyond the reach of all but large telescopes.   It has, however, two
little companions, distant about 30″.

60. Equuleus (Map IV.). — This little constellation is still
smaller than the Dolphin, and contains no such characteristic star
group.   It lies about 20° due east of Altair, and 10° S.E. of Delphinus
(see map).

Double Stars.   (1) ε, Mags. 5, 7.5; Pos. 73°; Dist. 11″.   The larger
star is also close double; Mags. 5.5 and 7; Pos. 290°; Dist. 0″.9.   Per-
haps resolvable with a 4-inch telescope.

61. Lacerta (Maps I. and IV.). — This is one of Hevelius's modern constellations, lying between Cygnus and Andromeda, with no
stars above the 4½ magnitude.   It contains a few telescopic objects,
but nothing suited to our purpose.

62. Pégăsus (not Pegas'us) (Map IV.). — This covers an immense space which is bounded on the north by Andromeda and Lacerta, on the west by Cygnus, Vulpecula, Delphinus, and Equuleus, on the south by Aquarius and Pisces, and on the east by Pisces and Andromeda. Its most notable configuration is " the great square," formed by the second-magnitude stars a, β, and γ Pegasi, in connection with a Andromedæ (sometimes lettered δ Pegasi) at its northeast corner. The stars of the square lie in the body of the horse, which has no hindquarters. The line drawn from a Andromedæ through a Pegasi, and produced about an equal distance, passes through ξ and ζ in the animal's neck, and reaches θ (third magnitude) in his ear. *Epsilon*, 8° northwest of θ, marks his nose. The forelegs are in the northwestern part of the constellation just east of Cygnus, and are marked, one of them by the stars η and π, the other by ι and κ.

Names. a, *Markab;* β, *Scheat;* γ, *Algenib;* ϵ, *Enif.*

Double Star. κ, Mags. 4, 11; Pos. 302°; Dist. 12''. The large star is also itself an extremely close double; Dist. 0''.3; (pointed at by the northern edge of the "square," at a distance one and a quarter times its length.)

Cluster. M. 15, A.R. 21ʰ 24ᵐ; Dec. 11° 38' (on the line from θ through ϵ, produced half its length, and just west of a sixth-magnitude star).

63. Aquarius (Map IV.). — This, the twelfth and last of the zodiacal constellations, extends more than $3\frac{1}{2}$ hours in right ascension, covering a considerable region which by rights ought to belong to Capricornus. It is bounded north by Delphinus, Equuleus, Pegasus, and Pisces; west by Aquila and Capricornus; south by Capricornus and Piscis Austrīnus, and east by Cetus. The most notable configuration is the little Y of third and fourth magnitude stars which marks the " water jar" from which Aquarius pours the stream that meanders down to the southeast and south for 30°, till it

reaches the Southern Fish. The middle of the Y is about 18°
south and west of α Pegasi, and lies almost exactly on the
celestial equator. A line drawn west and a little south from
γ (the westernmost star of the Y) to α Capricorni, passes
through β (third magnitude) at one-third of the way, and
through μ and ε (fourth and 3½) two-thirds of the way. α
(third magnitude) lies 4° west and a little north of γ. δ (third
magnitude) lies about half-way between the Y and Fomalhaut
in the Southern Fish, 3° or 4° east of the line that joins them.

Names. α, *Saad el Melik;* β, *Saad el Sund;* δ, *Skat.*

Double Stars. (1) ζ (the central star of the Y), Mags. 4, 4.5; Pos.
332°; Dist. 3″.6; pretty and easy. (2) 12 Aquarii (7° due west of β,
and the brightest star in the vicinity), Mags. 5.5, 8.5; Pos. 190°; Dist.
2″.8; yellowish white and light blue.

Clusters and Nebulæ. (1) M. 2; A.R. 21ʰ 17ᵐ; Dec. S. 1° 22′ (on
the line drawn from ζ through α, produced one and a quarter times its
own length). (2) II IV. 1, A.R. 20ʰ 58ᵐ; Dec. S. 11° 50′ (nearly on
the line from α through β, produced its own length, and 1¼° west of
ν; fifth magnitude); planetary nebula, bright and vividly *green.*

64. Piscis Austrinus (*or Australis*) (Map IV.).—This small
constellation, lying south of Aquarius and Capricornus, pre-
sents little of interest. It has one bright star, FOMALHAUT
(pronounced *Fomalhawt*), of the 1½ magnitude, which is easily
recognized from its being nearly on the same hour-circle with
the western edge of the great square of Pegasus, 45° to the
south of α, and solitary, having no star exceeding the fourth
magnitude within 15° or 20°. It contains no telescopic objects
available for our purpose.

South of it, barely rising above the southern horizon, lie the constellations of *Microscopium* and *Grus.* The former is of no account.
The latter is a conspicuous constellation in the southern hemisphere,
and its two brightest stars, α and β, of the second magnitude, rise
high enough to be seen in latitudes south of Washington. They lie
about 20° south and west of Fomalhaut.

LIST OF CONSTELLATIONS, SHOWING THEIR POSITION IN THE HEAVENS, AND THE NUMBER OF STARS IN EACH.

(The zodiacal constellations are denoted by italics, non-Ptolemaic constellations by an asterisk.)

R.A. h.	DECL. h.	+90° TO +50°	+50° TO +25°	+25° TO 0°	0° TO -25°	-25° TO -50°	-50° TO -90°
I,	II,	Cassiopeia, 46	Andromeda, 18; Triangulum, 5	*Pisces*, 18; *Aries*, 5	Cetus, 32	Phœnix, 32; *App. Sculp. 13	Phœnix, *bis.* 18; Hydrus, 13
III,	IV,	—	Perseus, 40	*Taurus*, 58	Eridanus, 64	(Eridanus, *bis.*)	*Horologium, 11; *Reticulum, 9
V,	VI,	*Cameloparidus, 36	Auriga, 35	Orion, 37; *Gemini*, 28	Lepus, 18	*Columba, 15	*Dorado, 16; *Pictor, 14; *Mons Mensæ, 12
VII,	VIII,	—	*Lynx, 29	Canis Minor, 6; *Cancer*, 15	Canis Major, 27; *Monoceros, 12	Argo-Navis, 133	Argo-Navis, *bis.* (Puppis) 9; *Piscis Volans,
IX,	X,	—	*Leo Minor, 15	*Leo*, 47	Hydra, 40; *Sextans, 3	—	Argo-Navis (Vela)
XI,	XII,	Ursa Major, 53	—	*Coma Ber. 20	Crater, 9; Corvus, 8	Centaurus, 54	Argo-Navis (Carina) 13; *Chameleon,
XIII,	XIV,	—	*Can. s Venat. 15; Boötes, 33	—	*Virgo*, 39	Lupus, 34	Centaurus, *bis.* 13; *Crux, 15; *Musca,
XV,	XVI,	Ursa Minor, 23	Corona Bor. 19; Hercules, 63	Serpens, 23	*Libra*, 22	Norma,* 14	*Circinus, 10

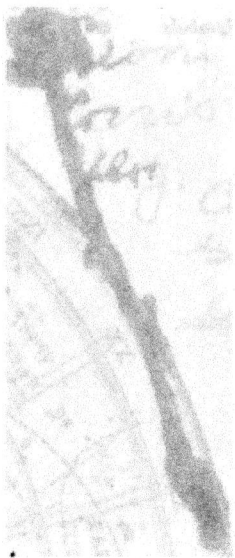

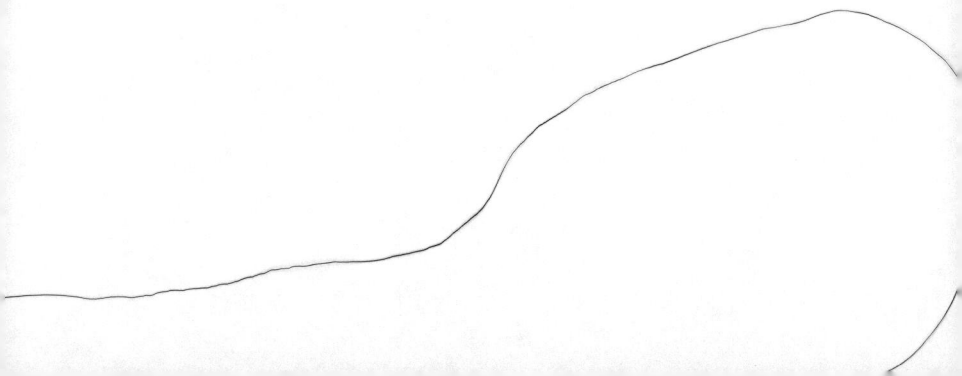

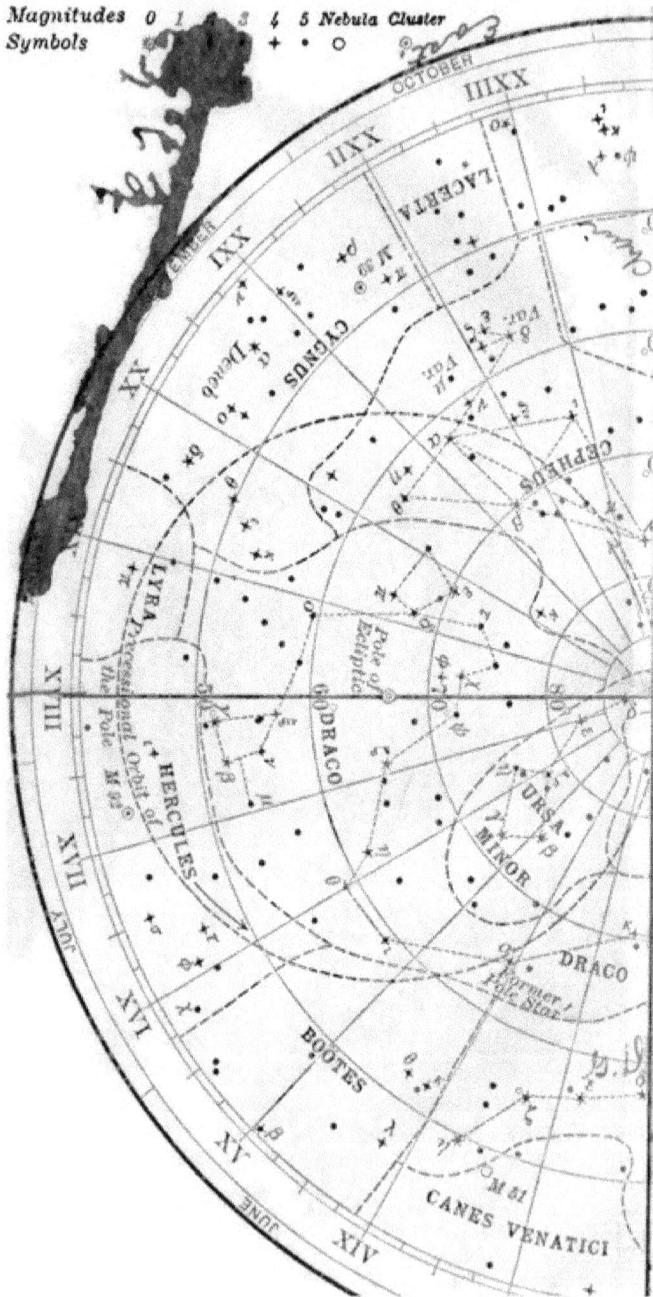

Magnitudes 0 1 2 3 4 5 Nebula Cluster
Symbols + • ○

OCTOBER

XXIII
XXII
LACERTA
XXI
CYGNUS
XX
CEPHEUS
Denb
LYRA
Precessional Orbit of
the Pole
M 92
HERCULES
DRACO
Pole of Ecliptic
URSA MINOR
DRACO
'Former'
Pole Star
BOOTES
M 81
CANES VENATICI
XIX
JULY
XVIII
XVII
XVI
JUNE
XV
XIV
XIII

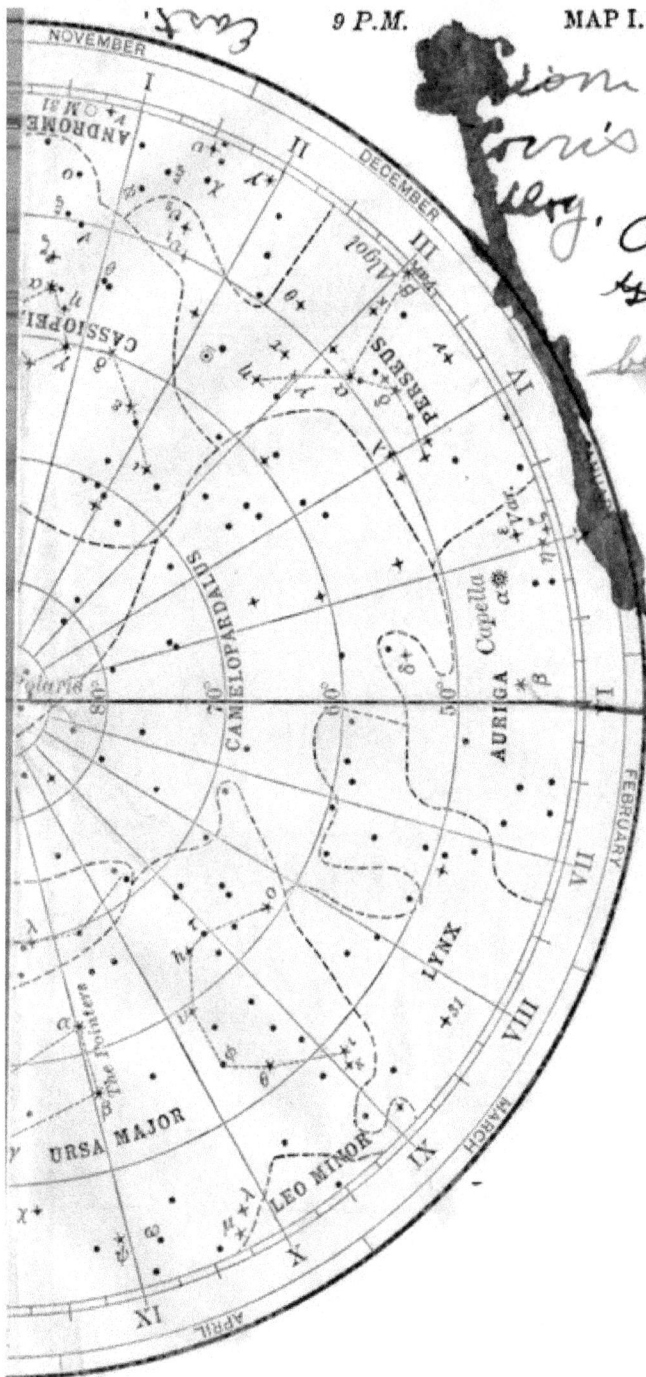

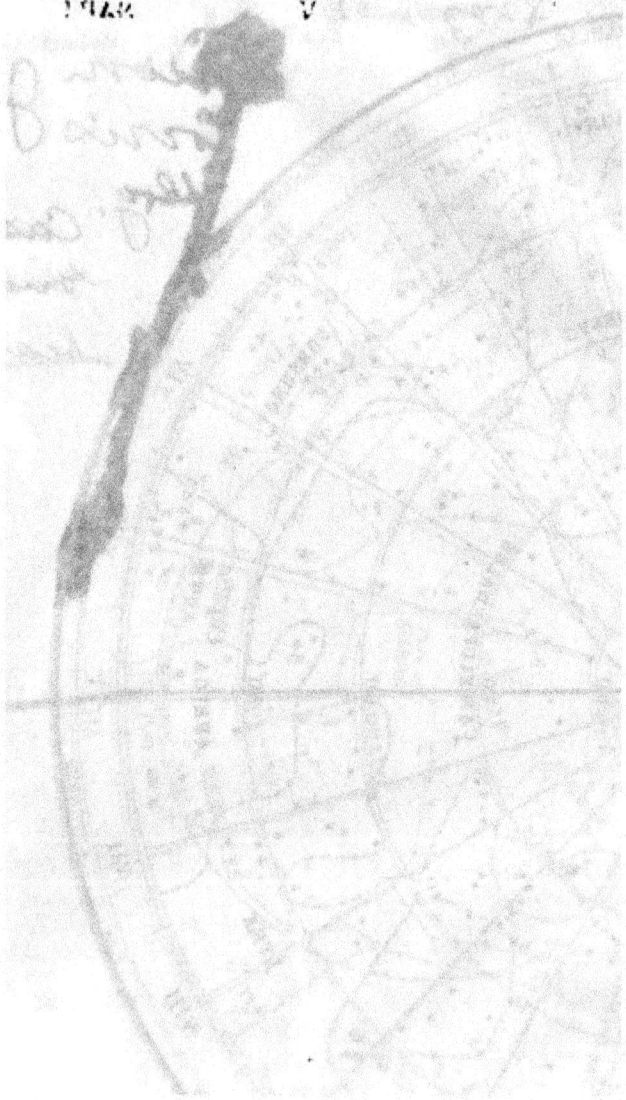

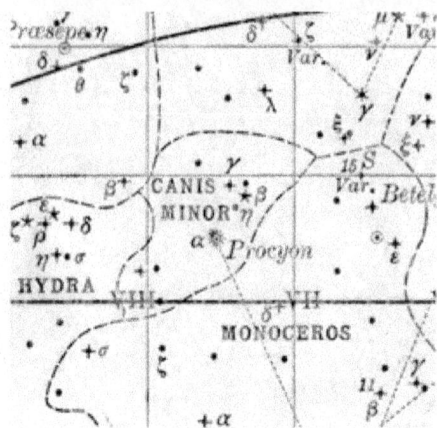

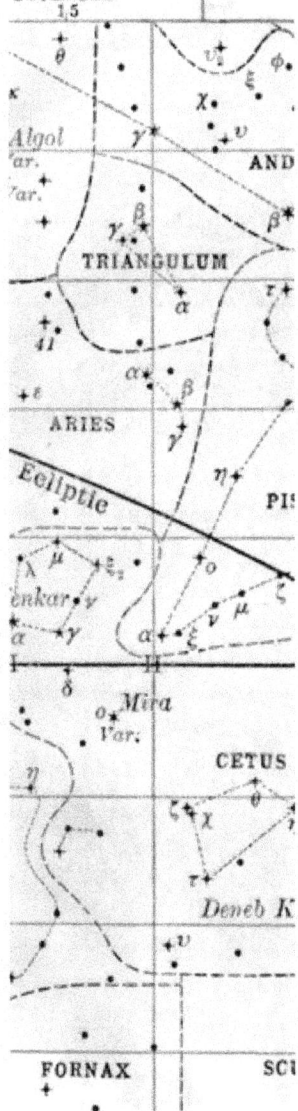

DECEMBER
15

θ

κ

Algol
Var.
Var.

γ

υ₂   φ

χ

ν

AND

β

β

γ

TRIANGULUM

α

τ

41

α

ε

β

ARIES

γ

η

Ecliptic

PIS

λ   μ   ξ₂

o

enkar   γ

ν   μ   ζ

α   γ

α   ξ

δ

o   Mira
Var.

η

CETUS

ζ   χ   θ

τ

Deneb K

υ

FORNAX

SC

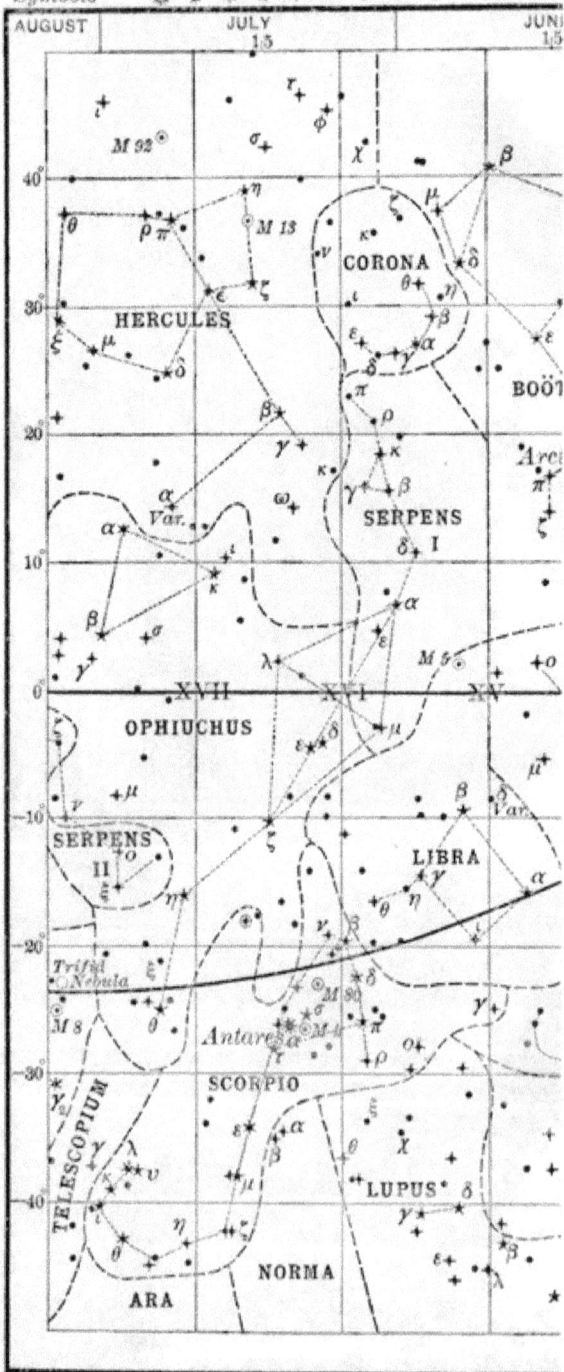

AUGUST | JULY 15 | JUNE 15

M 92 ◉

ι

σ

τ

φ

χ

β

40°

η

M 13 ◉

κ

μ

θ ρ π

ν

CORONA

δ

HERCULES

ζ

ξ

ι

θ

η

ε

β

BOÖT

ξ μ

δ

α

δ

ε

β

π

ρ

20°

γ

κ

κ

Arc

π

α
Var.

ω

γ

β

ζ

α

SERPENS

10°

ι

δ

I

κ

β

α

σ

ε

γ

λ

M 5 ◉

o

0° XVII   XVI   XV

OPHIUCHUS

ε δ

μ

μ

-10° ν

μ

β
Var

SERPENS

II o

LIBRA

γ

ξ

α

ξ

θ η

η

ν β

-20°

δ

Trifid
Nebula

ξ

σ

M 80 ◉

γ

M 8 ◉

θ

π

ρ

Antares ◉ M 19

τ

-30°

SCORPIO

ξ

γ

ε

α

χ

TELESCOPIUM

γ

β

θ

κ ν

μ

LUPUS δ

-40°

η

γ

θ

β

λ

NORMA

ARA

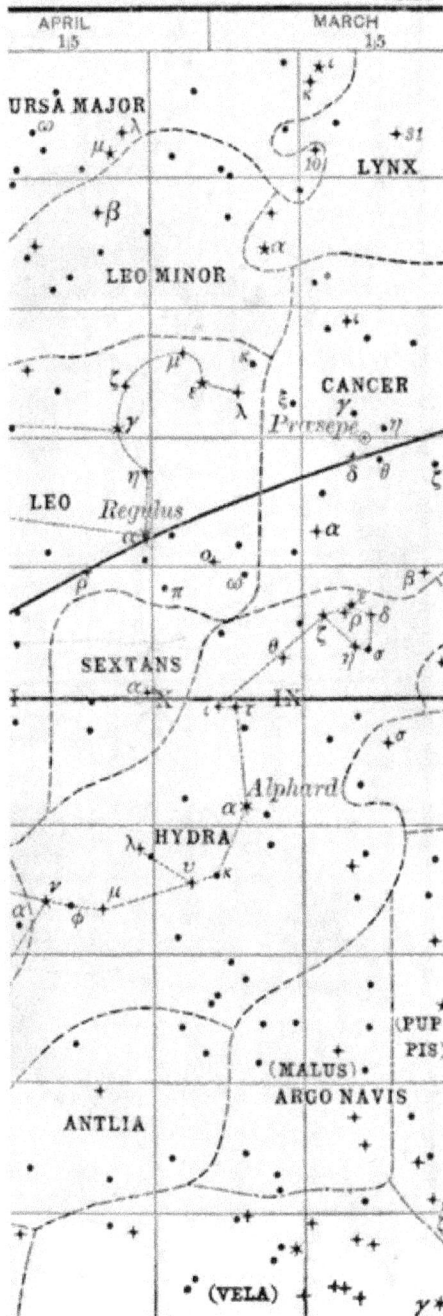

APRIL
15

MARCH
15

URSA MAJOR

ω
μ λ

+β

LEO MINOR

κ
31

10

LYNX

α

μ
ζ
ε
λ

κ

ξ
γ
Persepe

CANCER

η

η

LEO   Regulus
α

δ θ
ζ

α

ρ
π
ω

β

ε
ρ δ

θ
ζ
η σ

SEXTANS
α x
ι τ
IN
σ

Alphard

α

HYDRA
λ
υ κ
γ
μ
α
φ

(PUP-
PIS)

(MALUS)

ARGO NAVIS

ANTLIA

(VELA)

γ

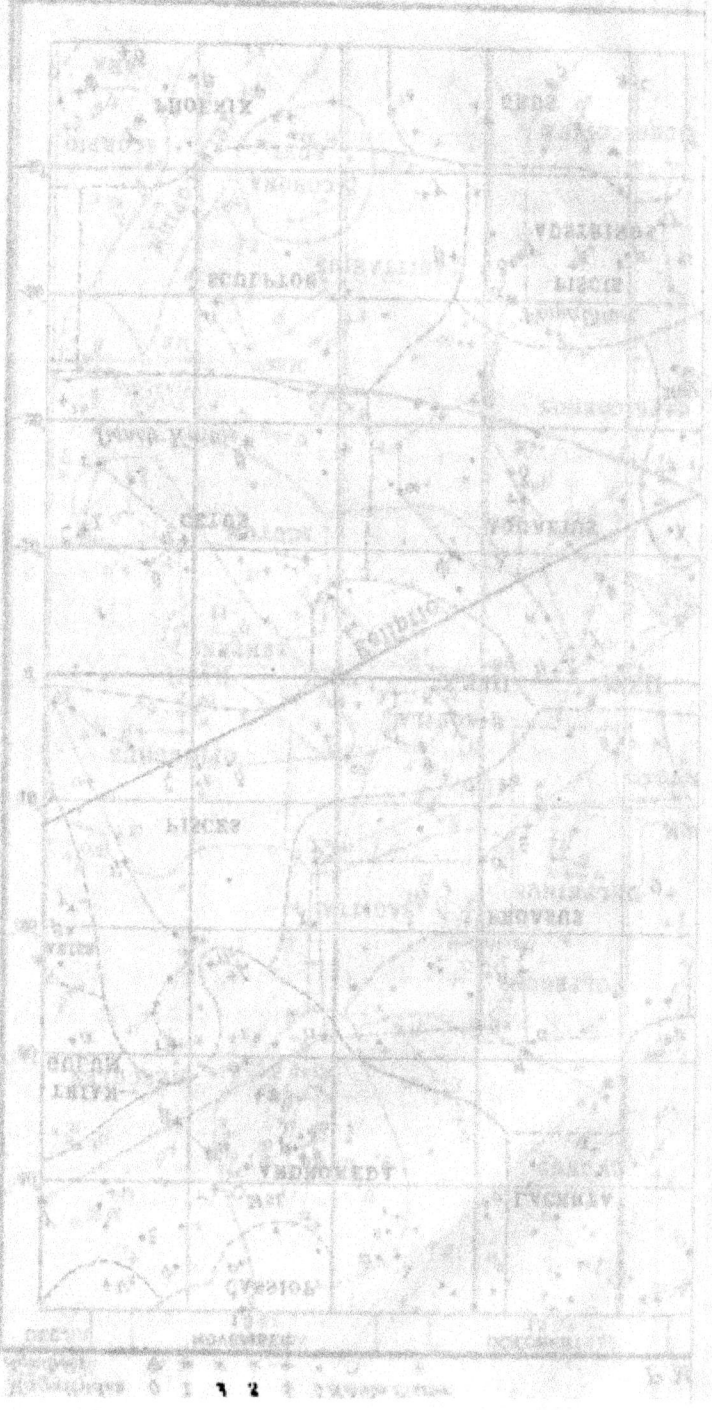

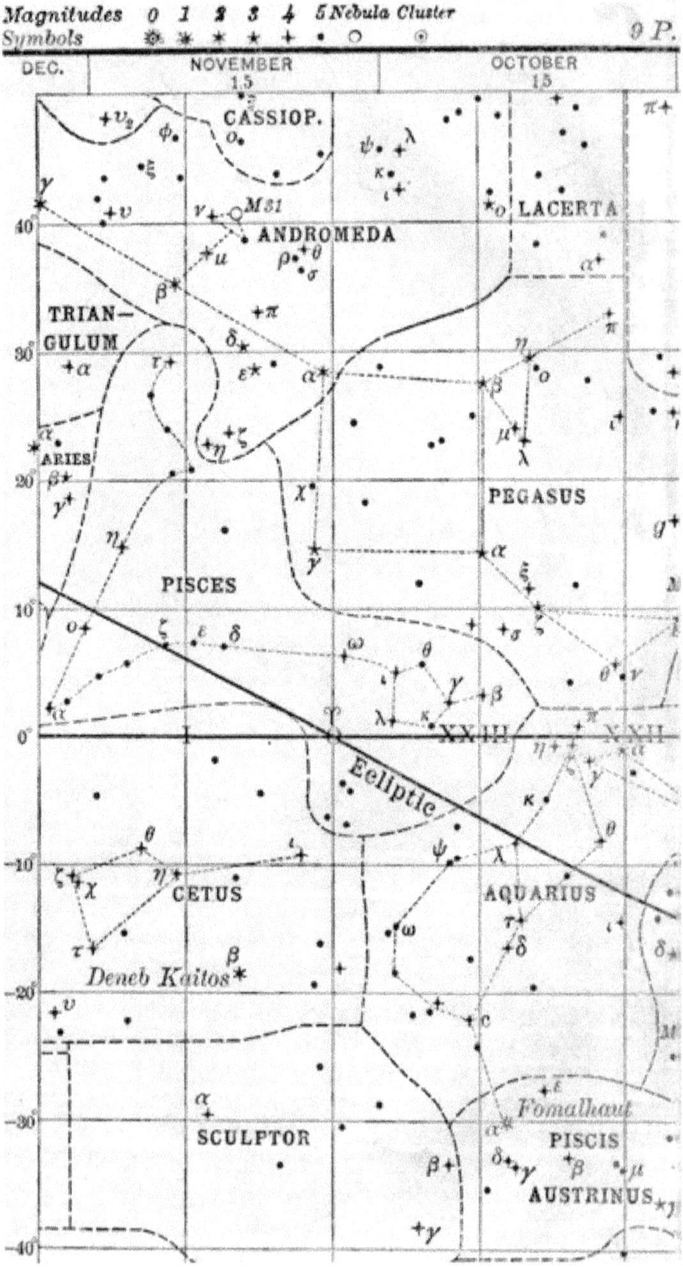

DEC. | NOVEMBER 15 | OCTOBER 15

π+

+υ₂   φ.   CASSIOP.   o.   ψ. +λ   κ   ι+

ξ   ν+ ○ M31   ψ. •λ   κ ι+   *o LACERTA

γ   +υ   ν+ ○ M31

40   ANDROMEDA   ρ.+θ   σ   α+

+u

TRIAN—   β✶   +π   η   π

GULUM   δ.✶   β+   o

30   +α   τ+   ε✶   α✶   μ+   λ

η+ ζ   λ

ARIES   +α   χ   PEGASUS

20   β+   α   ξ   g+

γ+   η+   α   ξ   M

η+   γ+   θ+ ν

10   PISCES   ω   θ   +σ

o+   ζ   ε   δ   ι   γ +β   θ+ ν

+α   λ+   κ   XXIII   η+   XXII   α

0   ℣   λ+ κ   η+   γ   α

Ecliptic   κ

θ   ι   ψ   χ+   θ

−10   ζ+ χ   η+   CETUS   ι   ψ   χ+   AQUARIUS

τ+   β   ω   τ+   δ   ι   δ✶

Deneb Kaitos ✶   +δ

−20   +υ   c   M

ε+   Fomalhaut

−30   α+   α✶   PISCIS

SCULPTOR   β+   δ+ γ +β •μ   AUSTRINUS ✶η

−40   +γ

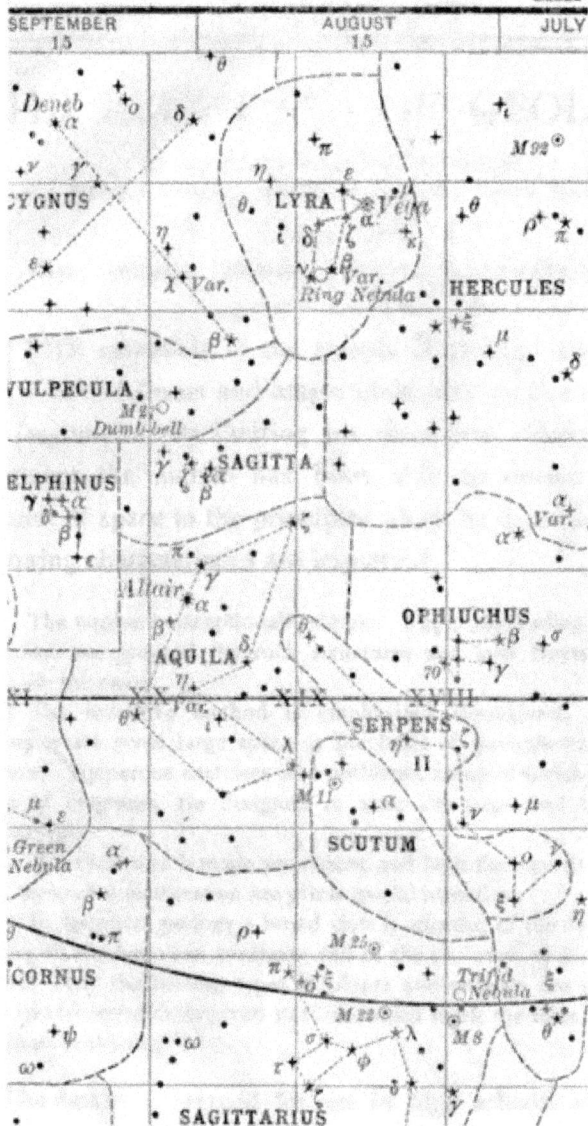

Deneb
α
o
δ
θ
π
M 92
ν
γ
η
ε
CYGNUS
θ
LYRA
ε
Vega
θ
ρ
π
ι
δ
α
κ
η
ζ
β
χ Var.
ν Var.
Ring Nebula
HERCULES
β
μ
δ
VULPECULA
M 2
Dumbbell
SAGITTA
γ ζ α
β
DELPHINUS
γ α
δ β
ε
π
α Var.
α
Altair
γ
α
OPHIUCHUS
β σ
β
AQUILA
δ
70
γ
η
Var.
XXI
XX
XIX
XVIII
θ
SERPENS
η
II
λ
M 11
γ μ
α
μ ε
o
Green Nebula
α
SCUTUM
ξ
β
η
π
ρ
M 25
CORNUS
π α ξ
Trifid Nebula
ξ
ψ
ω
M 22
σ λ
M 8
θ
ω
τ
φ
θ
γ
SAGITTARIUS

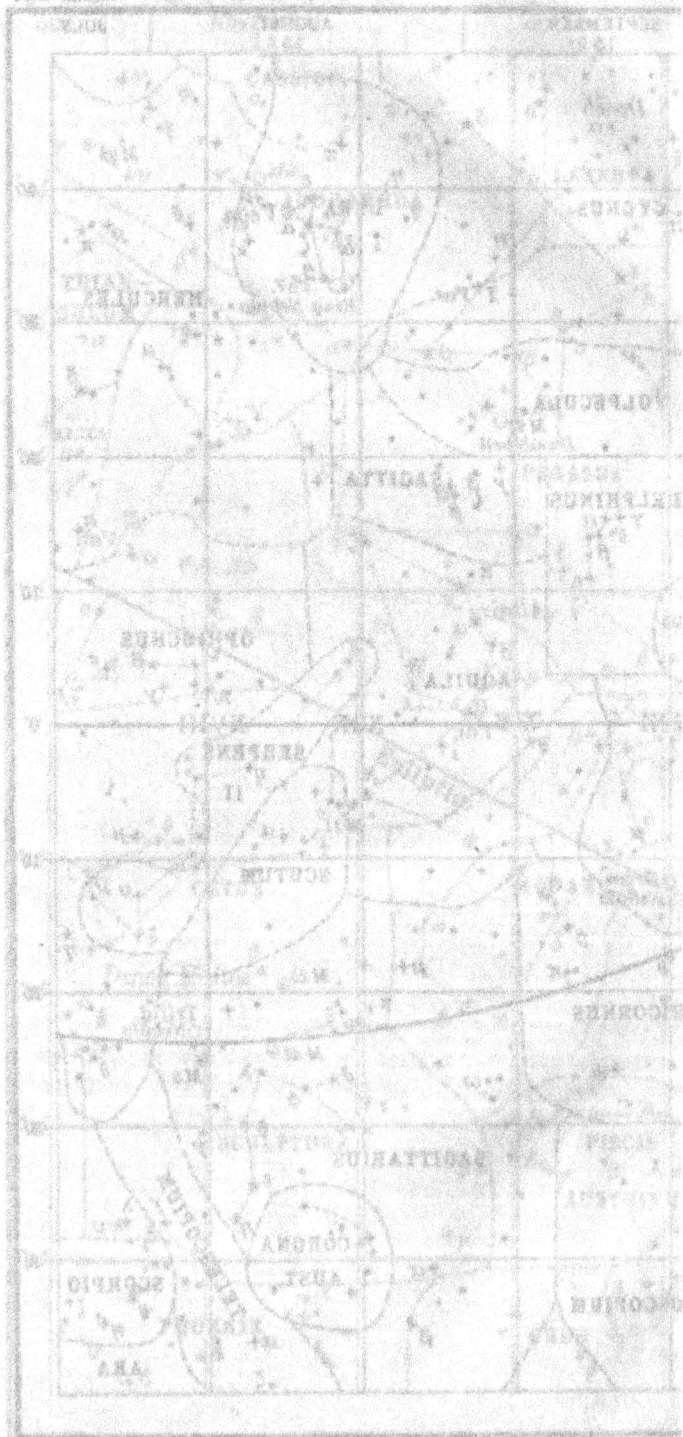

THE ELEMENTS OF GEOLOGY

By WILLIAM HARMON NORTON
Professor of Geology in Cornell College, Mt. Vernon, Iowa

8vo.   Cloth.   461 pages.   Illustrated.   List price, $1.40 ; mailing price, $1.55

THE essentials of the science of geology are treated with fullness and ample illustration in this text-book for beginners. By limiting his discussion chiefly to this continent the author has been able to devote a large amount of space to the principles which he describes. The following characteristics are important.

1. The outline is exceptionally simple. Under the leading geological processes are grouped the rock structures and land forms of which they are the cause.

2. The inductive method is emphasized throughout. Concrete examples are given large space as the basis of generalizations of the science. Numerous exercises and problems, many of which are in the form of diagrams, are designed to train the pupil and to test his knowledge.

3. The cycle idea is made prominent, and both the records of erosion and those of sedimentation are given special attention.

4. In historical geology a broad view is afforded of the development of the North American continent and of the evolution of life upon the earth. Only the leading types of plants and animals are mentioned, and special attention is given to those which mark the lines of descent of forms now living.

The book is designed for use in high schools and academies, and may also be found useful in short elementary college courses.

GINN & COMPANY PUBLISHERS

ELECTRICITY, SOUND, AND LIGHT

By ROBERT ANDREWS MILLIKAN, Associate Professor of Physics in
The University of Chicago, and JOHN MILLS, Instructor in Physics in Western Reserve University

8vo. Cloth. 389 pages. Illustrated. List price, $2.00; mailing price, $2.15

T HIS book represents a one-semester college course
which has been given in substantially its present form
for a number of years in The University of Chicago, Western
Reserve University, and several other similar institutions.

It is the outgrowth of the conviction that in courses of
intermediate grade in colleges, universities, and engineering
schools, a thorough grasp of the fundamental principles of
physics is not readily gained unless theory is presented
in immediate connection with related concrete laboratory
problems. It represents a complete logical development,
from the standpoint of theory as well as experiment, of the
subjects indicated in the title.

The book contains twenty-eight chapters, — sixteen in
electricity, five in sound, and seven in light. Each chapter
is concluded by the appropriate experiment. These experi-
ments have been put into such form that they demand no
special apparatus and can be performed by students in the
second year of college within the limited time of a two-hour
laboratory period. "Electricity, Sound, and Light" may be
combined with "Mechanics, Molecular Physics, and Heat,"
by Professor Millikan, to form the texts for an excellent
course of one year in college physics.

GINN AND COMPANY PUBLISHERS

YOUNG'S
MANUAL OF ASTRONOMY AND GENERAL ASTRONOMY

PROFESSOR YOUNG'S position among the great astronomers of the world is firmly established. As a clear and forcible writer he is equally well known. His series of astronomical text-books combines in an unusual degree the qualities of accurate scholarship, simplicity of style, and clearness of statement which belong to every successful text-book.

MANUAL OF ASTRONOMY

By CHARLES A. YOUNG, Professor of Astronomy in Princeton University. 8vo. Half leather. 611 pages. Illustrated. List price, $2.25; mailing price, $2.45.

Young's "Manual of Astronomy" is a new work prepared in response to a pressing demand from various quarters for a class-room text-book intermediate between the author's "General Astronomy" and his "Elements of Astronomy."

The subject-matter of the book has been derived largely from its predecessors, but everything has been carefully worked over, rearranged, rewritten where necessary, and added to in order to adapt it thoroughly to the end in view and to harmonize it with the latest astronomical research.

The unusually numerous illustrations are particularly noteworthy, both for their artistic excellence and for their value in explaining the text. Among them are a number of important photographs which have never before been inserted in a text-book. The mechanical execution of the book as a whole is above criticism, and has called forth the enthusiastic praise of the general reader, the scientist, and the teacher.

GENERAL ASTRONOMY

By CHARLES A. YOUNG, Professor of Astronomy in Princeton University. 8vo. Half morocco. 630 pages. Illustrated. List price, $2.75; mailing price, $3.00.

This text-book is intended for a general course in colleges and schools of science, and requires only an elementary knowledge of algebra, geometry, and trigonometry. It has been found an eminently satisfactory book for this grade of work.

GINN & COMPANY PUBLISHERS

TEXT-BOOKS ON ASTRONOMY

By **CHARLES A. YOUNG**

Late Professor of Astronomy in Princeton University

LESSONS IN ASTRONOMY. (Revised Edition.) Including Star Maps.
420 pages. Illustrated. List price, $1.25; mailing price, $1.40.

ELEMENTS OF ASTRONOMY. With a Uranography. 464 + 42 pages
aud four double-page star maps. List price, $1.60; mailing price, $1.75.

URANOGRAPHY. From the "Elements of Astronomy." Flexible covers. 42 pages
and four double-page star maps. List price, 30 cents; mailing price, 35 cents.

MANUAL OF ASTRONOMY. 611 pages. Illustrated. List price, $2.25;
mailing price, $2.45.

GENERAL ASTRONOMY. A text-book for colleges and scientific schools.
630 pages. Illustrated with 250 cuts and diagrams and supplemented with
the necessary tables. List price, $2.75; mailing price, $3.00.

A SERIES of text-books on astronomy for higher schools,
academies, and colleges, prepared by one of the most dis-
tinguished astronomers of the world, a popular lecturer and
a successful teacher.

The "Lessons in Astronomy" was prepared for schools that
desire a brief course free from mathematics. The book is fully
down to date, and several beautiful plates of astronomical objects
and instruments have been inserted in the revised edition.

The "Elements of Astronomy" is a text-book for advanced
high schools, seminaries, and brief courses in colleges generally.
Special attention has been paid to making all statements accurate.

The "Manual of Astronomy" is a new work prepared in response
to a pressing demand for a class-room text-book intermediate between the author's "General Astronomy" and his "Elements of
Astronomy." It is largely made up of material drawn from the
earlier books, but rearranged, rewritten when necessary, and added
to in order to suit it to its purpose and to bring it thoroughly
down to date.

The eminence of Professor Young as an original investigator in
astronomy, a lecturer and writer on the subject, and an instructor
in college classes, led the publishers to present the "General
Astronomy" with the highest confidence; and this confidence has
been fully justified by the event. It is conceded to be the best
astronomical text-book of its grade.

GINN & COMPANY PUBLISHERS